Patrick Proctor Alexander

Mill and Carlyle

An Examination of Mr. John Stuart Mill's Doctrine of Causation in Relation to Moral

Freedom

Patrick Proctor Alexander

Mill and Carlyle
An Examination of Mr. John Stuart Mill's Doctrine of Causation in Relation to Moral Freedom

ISBN/EAN: 9783337275136

Printed in Europe, USA, Canada, Australia, Japan

Cover: Foto ©berggeist007 / pixelio.de

More available books at **www.hansebooks.com**

Mill and Carlyle.

AN EXAMINATION

OF

MR JOHN STUART MILL'S

Doctrine of Causation in Relation to Moral Freedom.

WITH

AN OCCASIONAL DISCOURSE ON SAUERTEIG,

BY SMELFUNGUS.

BY

PATRICK PROCTOR ALEXANDER, A.M.

EDINBURGH:
WILLIAM P. NIMMO.
1866.

PREFACE.

THE papers on which the following Essay on Mr Mill is based, were written shortly after the appearance of his work on Sir W. Hamilton. They were written swiftly and recklessly for an immediate purpose, and on reading them as printed, I became aware that certain passages were conceived in a spirit of something which seemed like insolence, scarcely to be held becoming in relation to a man so eminent as Mr Mill. That in the process of revision and expansion, this original sin has been eradicated so entirely as might be wished, I cannot venture to be quite confident. But only a blockhead will imagine that in any little vivacities of expression I intend disrespect to Mr Mill, farther than as unable to profess respect for his reasonings on the topic under discussion. In the little Extravaganza which follows, surely I need not formally disclaim an offensive intention to Mr Carlyle, a man whom I entirely honour, and—though with only a modified belief in him as a prophet—consider simply our greatest man of letters now living. The thing was written merely *pour rire* on the appearance of the first two volumes of his "Frederick," and a few copies were printed for the amusement of a circle of friends. As the late conclusion of the work gives it anew a sort of pertinence, and under the mask of its wild fooling there are insinuated some morsels of not unserious criticism, I have thought it might bear reproduction. That Mr Carlyle himself—should it ever come under his eye—could see anything but matter of amusement in it—in so far as it may contain any genuine element of the amusing—it would truly surprise me to learn. The piece throughout abounds with glancing allusions which are only likely to be caught by readers almost critically familiar with nearly the entire round of Mr Carlyle's writings; but it does not seem worth while to indicate these—even if it could be done without trouble—inasmuch as a very cursory acquaintance with Mr Carlyle will quite enable a reader to appreciate in a general way such merit as the thing may be judged to possess.

P. P. A.

MR JOHN STUART MILL ON FREEDOM.

Though Hume, in the opening of his ingenious Essay, entitled, "Of Liberty and Necessity," confidently promised his readers, "at least some decision of a con-" troversy" to which, as it "turned merely upon words " and ambiguous expressions," " a few intelligible de-" finitions would immediately have put an end" any time for two thousand years previously, it has not been found that, since he wrote, unanimity of opinion exists among thinking men touching the points aforetime at issue. In our own day the old dispute re-emerges as frequently as ever before; and the writer, in the following remarks, has at least the excuse of them afforded by the fundamental importance and abiding interest of their subject. It is a subject not for philosophers only, but which all men professing themselves rational creatures are seriously concerned to meditate. It is impossible to think in any sense decisively on moral questions without being instantly led up to it; and for every man not content to be merely a piece of drift-wood on the seas of thought, borne hither and thither as the accident of the tides will, some at least provisional solution of the world-old problem it suggests is positively needed as a sort of intellectual *vade mecum*.

The following little Essay on the subject was suggested by a perusal of the chapter "On the Freedom of the "Will," in Mr Mill's late book on Hamilton, of the views set forth in which it is mainly an attempted refutation. These views are scarce in any respect different from those which have long been before the world in the chapter "Of Liberty and Necessity," as it stands in the successive editions of Mr Mill's "System of Logic." But inasmuch as in seeking to adjust them to the needs of his polemic against Hamilton, Mr Mill has here found it necessary to develop considerably the moral side of his argument, it seems reasonable to suppose he considers he has flooded the subject with new and important lights. As to this, we regret to be unable to agree with him. It seems to us he has left everything precisely as it was; except, indeed, as the confusions which very readily beset the inquiry appear in his pages rather worse-confounded than we have ever before chanced to see them. Mr Mill's chapter on Freedom—though in various influential quarters we have seen it selected for special laudation—is really the weakest in his book. The peculiarity of Mr Mill's view is this, that announcing unconditionally the Necessity of human actions, or—if he thinks it makes any difference, as oddly enough he seems to do—the Law of Causation as applied to them, he professes to maintain intact a system of Moral doctrine, which, except on the hypothesis of Freedom, is absolutely without a basis. It is our purpose to inspect his method of mixing his oil and water, and to show how little in the result, the dissentient fluids, even by a philosophical ingenuity so consummate as that of Mr Mill, can be coaxed into kindly interfusion. Within our proposed limits we cannot concern ourselves with the "accumulation of logical swim-"bladders"—as Mr Carlyle phrases it in his excruciating quiz of Coleridge—by means of which Sir W. Hamilton proposes to cross the metaphysical Jordan to the fair land of Freedom beyond. Nor is it highly essential we should. It was a good deal Sir William's way to pour his porter out with a somewhat high hand, and to pride himself—pretty much as we may see a waiter do—on the seething of scholastic froth which appeared as a head for the liquor. Quite ingenuously he seems to have considered that the complete philosophical soundness of his tap was in this way guaranteed. It is a notion not much countenanced by Mr Mill, who, with one puff of scornful breath incontinently blows off from the discussion a quantity of Doctrines of the Conditioned, opposite Inconceivables, Excluded Middles, and what not. Whether or no, as froth, Mr Mill has effectively made away with them, the disciples of Sir W. Hamilton may be left to inquire at their leisure. Our own easy notion in the matter is, that here, as throughout his book, Mr Mill has stormed the mere *outworks* of Hamilton's position with undeniable vigour and success; as instance, in his criticism of Hamilton's Necessity, considered as an Inconceivable, which we should a little dislike to have to answer. But has the citadel also fallen? Here, at least, it seems to us it has not. The Inconceivables, Middles, and so forth, being allowed to have fled upon the winds, we find, as the *residuum* of Hamilton's doctrine, our old friend the Moral Imperative—decisively announced by Kant, and as following him, by Coleridge—in order to belief in the validity of which, it is necessary to postulate Freedom. If, says Sir William, we be not directly conscious of Freedom, (according to Mr Mill, Sir W. does not quite know as to this,) we are at least conscious of Moral Responsibility, in which Freedom, as its ground is implied, and which—Freedom withdrawn—can be nothing but the merest figment. Mr Mill at this kernel of the dispute alleges in reply, that of Freedom we have no direct consciousness; and admitting our feeling of Moral Responsibility, he undertakes to find for it in his scheme of Necessity, that sound and satisfactory basis, the possibility of which, except on the hypothesis of Freedom, is by his opponents denied. It is the main object of this paper to inquire in how far he can be held to prosper in his attempt to harmonise our practical moral instincts with his speculative tenet of Necessity.

And first—can any real and important distinction be

made out between Mr Mill's Causationism—so call it—
and the understood doctrine of Necessity? Inasmuch
as Mr Mill, both here and in his "Logic," repeatedly insists on such a distinction as cardinal, some preliminary
inquiry as regards it seems called for. His is, he says,
a "falsely-called Doctrine of Necessity;" and he appends
to this a note as follows:—" Both Sir W. Hamilton and
" Mr Mansel sometimes call it by the fairer name of De-
" terminism. But both of them, when they come to
" close quarters with the doctrine, in general call it
" either Necessity, or less excusably, Fatalism. The
" truth is, that the assailants of the doctrine cannot do
" without the associations engendered by the double
" meaning of the word Necessity, which in this appli-
" cation signifies only invariability, but in its common
" employment, compulsion." Elsewhere we find him
writing—"If necessity means more than this abstract
" possibility of being foreseen; if it means any myste-
" rious compulsion apart from simple invariability of
" sequence, I deny it as strenuously as any one. To
" enforce this distinction was the principal object of the
" remarks which Mr Mansel has criticised. If an un-
" essential distinction from Mr Mansel's point of view,
" it is essential from mine, and of supreme importance
" in a practical aspect." The only other passage on
which Mr Mill relies for the establishment of this important distinction runs thus—" A volition is a moral
" effect which follows the corresponding moral causes as

"certainly and invariably as physical effects follow their physical causes. Whether it *must* do so, I acknowledge myself to be entirely ignorant, be the phenomenon moral or physical. All I know is that it always *does*."

Mr Mill seems in error in supposing that the argument against his doctrine can only attain an illusory success, by a sub-insinuation of *compulsion* in the use of the term Necessity. All that is really needed for the perfect validity of that argument is, Necessity negatively defined as contradictory and exclusive of Freedom; and Mr Mill's Causationism is, on his own showing, with such a scheme of Necessity identical.) As to whether an act *not free*, can accurately be said to be *compelled*, this is an outlying question, the consideration of which may be postponed; enough, meantime that acts are only assumed necessary in the sense of their not being free. According to Mr Mill, when he sees a stone unattached fall to the earth, he simply knows it *does* fall, not that it *must*, or does *necessarily* fall. That stones certainly and invariably *do* fall, and that any particular stone *will*, under given circumstances, fall, and cannot, unless a miracle were wrought to prevent it, be rationally conceived of as doing otherwise, is readily by Mr Mill admitted; but, that it *must* fall, Mr Mill peremptorily declines to admit. Will he admit the stone powerless to prevent its fall? It may reasonably seem he ought to do so; everywhere else in the universe, denying—or more accurately, declining to admit any valid inference of—*power*, he can scarcely allege it in the stone. And if the stone be admitted passive and helpless in the matter, all as regards it is admitted that any mortal need care to contend for as meant by Necessity, determining human action. The defect in man of any power to act otherwise than as he does act, in other phraseology, of Freedom—is with some show of reason asserted to annihilate him as a moral and responsible agent. For that which a man is utterly unable to *help* doing, it is held absurd to impute to him either *praise* or *blame*. This is, in effect, the moral argument for Free-will, as against Mr Mill's Causationism; and plainly it rests securely enough on the mere negative assumption of impotence, as distinct from any positive compulsion, either asserted or implied.

Whether, again, a defect of any power to control it in the subject whereon the effect is operated, may not be the logical equivalent of a power of compulsion in the cause, it may be worth while to inquire. It may facilitate the decision of this question to substitute for the stone and its fall, in which compulsion is held inadmissible, a case in which it will not, by plain men at least, be denied. Suppose then, ten big men thewed like Hercules to clutch hold of a small and weak one, and *per* force drag him after them,—is there for Mr Mill in *this* case, any *must*, or inference of Necessity? If Mr Mill, like a mere man of common sense, decides to answer—Yes,—he implicitly throws up his brief; he admits here a *must* and a Necessity, which elsewhere, having made this admission, he will in vain seek to deny; for that this and every other conceivable case of compulsion admit of being generalised under Mr Mill's law of Causation, defined as simply "invariable se-" quence," is too obvious to be more than merely suggested. Should Mr Mill on the other hand decide, as in the case of the stone, to answer as becomes a philosopher of his school, that he can admit no *must* in the matter; that when the weak captive is swept away by his ten captors, we are not entitled to say, he *of necessity* goes with them; that what we are sure of is, that he *does* go, and always *will* go, but that any *power* in the ten muscular giants to *compel* his going, we have no right to assume; his deliverance would perhaps be profound, but we own we should find it puzzling. It is a popular error, it seems, to suppose *compulsion* in any case made out. It is, however, an error in which Mr Mill himself so far shares, that in any such case of *apparent* compulsion as that given, he would admit the subject of the outrage annihilated for the time being as a morally responsible agent. Of this, there can be no doubt, for we shall find him writing thus,—" Yes—if he ' could not " ' help' acting as he did—that is, if his *will* could not " have helped it; if he was under physical constraint," in which case Mr Mill, in common with the mass of men not philosophers, concedes "exemption" from blame and just penalty. Causation, therefore, in such an instance of it as that specified, admittedly involves, if not a *must*, Necessity, or compulsion, some such equivalent or *analogue* of these—Mr Mill may give it a name at his leisure—as serves to obliterate Responsibility, and nullify moral judgments. And, if in one case of Causation, this nameless equivalent of compulsion is present, we may fairly ask Mr Mill to show ground of its exclusion in others. Unless Mr Mill is prepared to announce one doctrine of Causation for gentlemen under constraint, and another for gentlemen at large, stones and the like inanimate bodies, he must needs confess his distinction between the doctrines of Causation and Necessity, in relation to the moral problem, a trivial and merely verbal one.

And in truth, though Mr Mill, as we saw, very much insists on this distinction, as it may seem to suit the exigency of his argument, it might almost appear that, apart from this, he does not habitually define it to himself with any great rigour or precision. Thus, we shall find him writing of "people not being punished for "what they were *compelled* to do," and this not by physical violence, but under urgency of some such motive as a fear of instant death. Nay, more; we shall find him identifying this moral compulsion with compulsion by physical constraint, in so far as to admit or assert that in neither case "could the *will* of the man "have *helped*" his action. Mr Mill thus explicitly admits as involved in the idea of Causation, physical alike and moral, the very *compulsion*, against the alleged surreptitious implication of which by his opponents, in calling his a doctrine of Necessity, we have seen his repeated protest. It is open, of course, to Mr Mill to say, that he uses here the word "compelled," merely in a loose and popular way, and really means something quite different; but people so trained as Mr Mill, to accuracy in the use of terms, will seldom in philosophical discussion, say things which they do not mean, except when their meaning is to themselves a little indistinct. Mr Mill would scarcely here have used the word *compelled*, unless he had been wont more or less— when its relation to his argument ceased to be before his mind—to associate the ideas of compulsion and Causal sequence, in some such way as to neutralise his distinction between the doctrines of Causation and Necessity. Nevertheless, supposing the distinction valid, we shall, in all that follows, use the convenient words,— Necessary, Necessity, &c., simply as implying the absence of Freedom, which Mr Mill in his Causationism maintains, and not as in any case including the element of positive compulsion he denies in it.

Following Mr Mill in his attempt to resolve the difficulties which beset this question, by representing them as originated and maintained by the use of inappropriate terms, Mr Bain (see "The Emotions and the Will," p. 544,) not only censures as "obnoxious" in this relation

the word "Necessity," but objects to "Freedom" as equally so. "One answer," he writes, "to be made to
"the advocates of Free-will, is, I conceive, the utter
"inappropriateness of the name, or notion, to express
"the phenomenon in question. We may produce any
"amount of mystery, incomprehensibility, insolubility,
"transcendentalism, by insisting on keeping up a
"phraseology, or a theoretical representation that is
"unadapted to the facts. I can imagine some votary of
"the notion that polar force (as in the magnet) is the
"type and essence of all the powers of nature, finding
"the difficulty of bringing gravity under it, and there-
"fore declaring the case of gravity an insoluble problem.
"In like manner, I believe that to demand that our
"volitions shall be stated as either free or not free, is
"to mystify and embroil the real case, and to superadd
"factitious difficulties to a problem not in its own na-
"ture insoluble. Under a certain motive, as hunger,
"I act in a certain way, taking the food that is before
"me, going where I shall be fed, or performing some
"other preliminary condition. The sequence is simple
"and clear when so expressed; bring in the idea of
"Freedom, and there is instantly a chaos, imbroglio,
"or jumble. What is to be said therefore, is that this
"idea ought never to have come into the theoretical
"explanation of the Will, and ought now to be sum-
"marily expelled. The term 'Ability' is innocent, and
"has intelligible meanings, but the term Liberty (or

"Freedom) is brought in by main force into a pheno-menon to which it is altogether incommensurable." Its introduction, Mr Bain proceeds to censure as "the conversion of metaphor into scientific language," and then concludes of it thus—"We understand the difference between slavery and free citizenship, between a censorship and a free press, and between despotism in any shape and the liberty of the subject; but, if any one asks whether the course of a volition in a man or an animal is a case of despotism or a case of freedom, I answer that the terms have no relation whatsoever to the subject. The question put into some one's mouth by Carlyle, 'Is virtue then a gas?' is not too ridiculous a parody upon the foregoing."

Of all this, what is to be said? Simply that there is nothing whatever in it. The use of the word Freedom in regard of the Will is indeed metaphorical; but the metaphor employed is so close and apposite that Mr Bain's is probably the only mind that ever saw in it a possible source of confusion. Let us ask any one using the term what he means by Free-will, and we are at a loss to know how he could define it except as a power or *ability* in man at any moment, to act otherwise than as he does act. If we "summarily expel," as Mr Bain desires, the term "Freedom" which is so "obnoxious," as instantly leading to a "chaos, imbroglio, or jumble," and substitute for it the term "Ability," which, it seems, is "innocent and intelligible," the argument in every iota

of it remains precisely as it was. If there be no Freedom
in man, no *ability* in him, that is, (it would require some
little ingenuity to explain the term save thus,) to act
otherwise than as he does act, his *inability* to do so
must be held to be as utter and absolute as the inability
to act at all of a man tied tight with cart ropes and
flung on the ground. How then is he in reason to be
held criminal in not having acted otherwise, any more
than the man in bonds for not having acted at all, supposing him under an obligation to act, if *free?* Do the
bonds of the one man in any sense more rigorously
determine his inaction, than the causal motive determines
the action of the other man, and incapacitates him from
acting otherwise? Surely it will not so be held by either
Mr Mill or Mr Bain. The *inability* of a man to act
in anything except as he does act, is then as complete
as if in his act he were *compelled*. The antithesis between physical constraint and freedom may thus be
logically identified with that between Causal urgency of
motive and an ability in man to act otherwise than as he
does act. It is a metaphor as used; but the only minds
in which it ever yet led to any confusion worth speaking
of are the minds of the philosophers, who, following
Hume, the originator of this whole line of argument,
have thought that by extruding it as a metaphor they
got rid of any of the difficulties which are really essential
to the subject. Moreover, till a case of physical constraint
or compulsion be produced, which is not also a case of

Causation defined as "invariable sequence," there are tolerable grounds for elevating it from the rank of a mere metaphor to that of an illustrative instance. But, as we said above—Necessity being objected to by Mr Mill as implying compulsion—that we should not so use the term, but merely as implying the negation of Freedom—so now—the term Freedom being by Mr Bain objected to—we shall never in what follows use *it* except in his severely "innocent" sense of an "Ability" in man to act otherwise than as he does act. By Necessity we shall be bound to mean throughout simply Causation, or constant and unconditional sequence; by Freedom an Ability in man as stated. And it is our hope it will be seen that, by these concessions to opponents, the force—such as it ever may have been—of the moral argument in favour of said Freedom or Ability is touched in no jot or tittle.

The "direct consciousness of Freedom," asserted by Hamilton, as Mr Mill alleges, "in a doubtful and "hesitating manner," but "by many maintained with "a confidence far greater than his," Mr Mill distinctly denies. As this supposed doubt and hesitation in Hamilton is inferred by Mr Mill from certain slight apparent discrepancies in his statements given at different times, it seems more or less pertinent to note that Mr Mill, in his own statements of the matter, is by no means so consistent as might be wished. In his "Logic" we find him writing—"The metaphysical theory of

"Free-will (*for the practical feeling of it common in a greater or less degree to all mankind*, is no way inconsistent with the contrary theory,) was invented," &c.,—and again—"We shall find that this feeling of being able to modify our own character, *if we wish*, is itself *the feeling of moral freedom we are conscious of*." It will, we hope, be found proved at a later stage of the discussion, that this of our "being able to modify our own character, *if we wish*," is a use of words without meaning, unless some admission of Freedom be implied in it; and even were it not so proved, the latter clause of the sentence retains its full significance. We are thus entitled to say that Mr Mill *had* at one time a consciousness of Freedom; now he assures us he has it not. The explanation of this is probably to be found in such passages of his later book as the following—"All agree with him (Hamilton) in the position that a real fact of consciousness cannot be doubted or denied."—"Consciousness, it will probably be said, is the best evidence; and so it would be, if we were always certain what is consciousness,"—and quotations might at will be accumulated to the like effect, that a *datum* of consciousness, if genuine, must be held a deliverance of truth. Whilst a practical feeling or consciousness was held by Mr Mill un-authoritative, as "no way inconsistent with a contrary theory," Mr Mill had a consciousness of Freedom; now that a closer contact with Hamilton has forced on him the authority of consciousness, the denial of which would leave Science itself without a basis, we find that his consciousness of Freedom has departed. Might it not almost seem to have departed in the interest of "the contrary theory?" Be this as it may, Mr Mill, without any of the doubt and hesitation ascribed to Hamilton—some *modicum* of which might not have been amiss on his own part, the state of the case considered—is here found "rejecting as "a figment" the consciousness of Freedom, which elsewhere he frankly admits; and though his statement of the matter cannot, perhaps, on its own ground, conclusively be shown to be erroneous, it does not seem a hopeless task somewhat to reduce its force as against the advocate of Free-will doctrine. "But this conviction," writes Mr Mill, "whether termed consciousness or only " belief, that our will is free—what is it? Of what are we " convinced? I am told that whether I decide to do or to " abstain, I feel that I could have decided the other way. " I ask my consciousness what I do feel, and I find indeed " that I feel (or am convinced) that I could have chosen " the other course, *if I had preferred it* (sic); but not " that I could have chosen one course while I preferred " another." Of this it seems enough to say that, as Mr Mill proceeds to define what he means by *preference*, as the final award, elective act, or *choice* of the mind on a view of the whole circumstances, the distinction which he here makes is a distinction without a difference. The advocate of Freedom has only to reply—as it is plainly

competent for him to do—that this final award of the mind, elective act, choice, or—if Mr Mill pleases—*preference*, which is scarce, except nominally, to be distinguished from the external act in which it issues, is the very thing over which he is conscious of a free exercise of power. Feeling he has amiss *preferred* to *do*, he now also feels he *ought* to have *preferred* to *abstain*, as he further feels that he could have done. Clearly Mr Mill makes no way whatever here; let us follow him a little further. " Take any alternative, say to murder or not to
" murder. I am told that if I elect to murder, I am con-
" scious that I could have elected to abstain; but am I
" conscious that I could have abstained, if my aversion to
" the crime, and my dread of its consequences, had been
" weaker than the temptation? If I elect to abstain, in
" what sense am I conscious that I could have elected to
" commit the crime? Only if I had desired to commit it
" with a desire stronger than my horror of murder, not
" with one less strong." Now, had all our " desires and
" aversions" (motives) been *passions*—using the word in its primary and accurate sense*—such as the blind, inevitable animal rage of thirst or of hunger, or the physical shrinking from a red-hot iron of one whose flesh has been sometime seared with it, Mr Mill's reasoning† here would

* As here,—" An Agent over-ruled by a blind impulse is a contradiction in terms; for then he is not an Agent at all, but a mere *Patient*."
—Dr Samuel Clarke.

† That after all, it is *reasoning*, and reasoning a little out of place, will presently fall to be noted.

B

have been stronger than any chain cable. But that the fact is far otherwise, Mr Mill himself will perhaps not dream of denying. The great mass of our desires and aversions are as truly and efficiently *acts* in which the will is immanent—and by consequence *free* acts, in so far as the will may be free—as the external acts in which they issue. Even within the range of animal appetite, this might perhaps in a degree be made obvious. The "desire to commit murder," again, is as accurately a moral *act*, as is the murder itself which comes of it. Moral "desires and aversions," more particularly, are indeed *motives*, but they are also—and antecedently in the order of logic—*acts*, as the rage of a hungry man for food cannot be. If this distinction in the least be valid, very obviously, in his neglect of it, Mr Mill's fabric of reasoning goes to pieces like a child's card castle. For, if motives be also acts; and *free* acts—as provisionally for the purpose of the argument we are plainly entitled to assume them—the resulting acts are accurately to be termed free, however inevitably determined by *their* determinations. Include, from this point of view, in the consciousness of Freedom, not simply the act, but the act with its motive, considered as one complex phenomenon, and the reply to Mr Mill seems sufficient. Of course it is simply as such it is offered, not the least as a *rationale* of Freedom, which remains an utter mystery as before. For the motive, considered as an act, must depend on some previous motive, by which it in turn was determined; and so through a regressive series, in which Freedom fleets for ever one step back from us, and is never to be caught and detained. But it may be that the distinction indicated between mere blind animal impulse, and other "desires and aversions," Mr Mill would decline to admit. It would be pleasant to be assured that he would, inasmuch as in virtue of his doing so, we should be spared a good many further pages of more or less perplexed discussion. As thus—Nobody not idiotic would censure a man for *hunger*, or consider him morally responsible for the amount or urgency of the appetite; so that plainly, if we include all other desires and aversions under a common law with it, there is an end of our Moral Responsibility. Surely it will not be alleged that though the appetite itself must be held blameless, its lawless indulgence—as by theft, let us say —is not so. For the restraining desires and aversions (motives) are by the hypothesis included under the same blind law of irresponsibility; and we are no more answerable for their deficiency as check, than for the excess of the appetite as impulse. Obviously, therefore, there can be no more a Moral Responsibility in man, than we surmise it in a balance on which weights tilt each other up and down. Morally, he *is* just such a dead, irresponsible balance, on which some deity or devil, experimenting *in corpore vili*, weighs motives of desire and aversion. And precisely this inference from his doctrine it is, which, as pressed upon him by the advocates of

Freedom, the whole subsequent argument of Mr Mill is a hopeless struggle to evade.

Again, to quote Mr Mill,—" When we think of our-"selves hypothetically as having acted otherwise than we "did, we always suppose a difference in the antecedents; "we picture ourselves as having known something that "we did not know, or not known something that we did "know, which is a difference in the external motives, or as "having desired something or disliked something, more or "less than we did, which is a difference in the internal "motives." Setting aside here the so-called "external "motives"—which are *not* in accuracy motives, but merely* external indeterminate novel *conditions* of motive, to import which into the discussion is only by so much to embarrass it—in order to deprive Mr Mill's argument of all force, we have only to repeat of the "internal motives," that besides being motives, they are *acts*, of which we feel we might have determined the differences indicated. "When we think of ourselves "hypothetically as having acted otherwise than we did," we do indeed "suppose a difference in the antecedents;" but along with the new antecedents given in thought, there is given a conviction, that it was in our power to have generated them in act, and thus to have determined the result differently.

---

* "A motive is a desire or aversion"—page 519—an "external mo-"tive" is thus plainly an inaccurate synonym of the object of desire, or external condition of motive.

Previous to choosing one of two courses of conduct, we seem unquestionably to have a consciousness or conviction that it is in our power to choose either; in the intimate moment of act, this consciousness of needs disappears, but only to reinstate itself afterwards in a sense that it was in our power to have acted otherwise. And if this be not an authentic consciousness of Freedom, we see not in what terms it would be possible to define such a consciousness.

We have thus in consciousness two distinct testimonies to our Freedom: the testimony anterior to act, and that which follows it. Plainly, in the constant correlation of these—in the ratification of the conviction as it first shows itself by its subsequent reappearance—of the re-emergent conviction by that which has preceded it—lies the force of the affirmative assertion; and no negative which does not conclusively resolve both can be held as of weight against either. Mr Mill's attempted reduction of the consciousness of Freedom, in regard of acts which are past, our readers are so far in a position to estimate; to enable them to estimate likewise the success of his attempt to reduce our consciousness of Freedom in acts contingent and meditated, let us quote the sum of his wisdom on the subject, as conveyed in the following passage :—

"To be conscious of Freedom must mean to be con-
" scious before I have decided, that I am able to decide
" either way. Exception may be taken *in limine* to the

"use of the word consciousness in such an application.
"Consciousness tells me what I do or feel. But what I
"am *able* to do is not a subject of consciousness. Con-
"sciousness is not prophetic; we are conscious of what
"is, not of what will or can be. We never know that we
"are able to do a thing except from having done it, or
"something equal or similar to it. We should not know
"that we are capable of action at all if we had never
"acted. Having acted, we know, as far as that experience
"reaches, how we are able to act; and this knowledge,
"when it has become familiar, is often confounded with,
"and called by, the name of consciousness. But it does
"not derive any increase of authority from being mis-
"named; its truth is not supreme over, but depends on,
"experience. If our so-called consciousness is not borne
"out by experience, it is a delusion. It has no title to
"credence but as an interpretation of experience; and if
"it is a false interpretation, it must give way."

One is grieved to find a man like Mr Mill so beneath himself as here he must be held to be. "Consciousness "tells me what I do or feel. But what I am *able* to do "is not a subject of consciousness." Perhaps it is not; but what I *feel* I am able to do is surely a subject of consciousness; certain it is at least, it was at one time by Mr Mill himself so considered—*vide* "System of "Logic," as before quoted—"The practical *feeling* of "Free-will common in a greater or less degree to all man-"kind." "The *feeling* of moral Freedom we are *conscious* "of." And as Mr Mill himself now interprets this feeling of Freedom of which he was at one time conscious, it "must have meant" a being "conscious before he "had decided that he was *able* to decide either way." Yet the next instant we are told that "what I am *able* to do "is not a subject of consciousness." Mr Mill is here in the sort of confusion which he would probably have been a little severe upon had he met with it in Sir W. Hamilton. As to "consciousness is not prophetic; we "are conscious of what is, not of what will or can be," it seems enough to say that if we are conscious of a free force of volition continuously inherent in us, we are conscious of what *is*. And this is perhaps the best way of putting it; for it seems that we must not speak of being conscious of a free ability to act, or indeed of any such ability. According to Mr Mill, we *know* that we are able to act, but have no *consciousness* of being able, though how this should be, unless knowledge is denied as a mode of consciousness, Mr Mill has omitted to explain. In brief, this feeling or consciousness of being able of two ways to decide in either, which Mr Mill aforetime admitted, he does not, and cannot now deny. But as consciousness is now authoritative, as then it was not so decisively seen to be, he denies it as a consciousness; and merely that his "theory" may be kept intact, degrades it into a "*so-called* consciousness," and as such, a source of "delusion,"—"If our so-called con- "sciousness (of being 'able to decide either way') is not "borne out by experience, it is a delusion. It has no title to credence but as an interpretation of experience; and if it is a false interpretation, it must give way." From this we must infer Mr Mill to mean that the fact of a man's acting in *one* way is a satisfactory contradiction by experience, of his previous consciousness or conviction, that of two ways, he had power to choose *either*. But it is plainly nothing of the kind; such a consciousness as that in question experience can neither confirm nor invalidate. What experience assures us of is that the man *did* act *so;* as to whether he could or could not have acted *otherwise*, experience can tell us nothing. Would Mr Mill bethink him of his own principles? Of the fall of a stone Mr Mill will say,—All that I know is that it *does* fall, not that it *must*. By what right does he now imply such a *must* in the action of the man, as entitles him with absolute dogmatism to say that he *could not* have acted otherwise,—that experience has shown that he could not? It is sufficiently obvious that, from "experience," beyond what the man *does*, Mr Mill can know nothing of the matter. Wherefore, what he really means must be held to be, that everything which conflicts with his private little theories, is without further necessity of proof to that effect, discredited as mere "delusion."

But after all, a question of consciousness is a question of fact, not of argument, and must be decided by the simple appeal to consciousness. The only real question in the matter is to whose consciousness the appeal is to be made? And the answer to this must plainly be,— To the general consciousness of the race, *philosophers with rigour excepted.* The necessity of this exclusion is obvious; inasmuch as we may be sure no philosopher will ever find anything in his consciousness which would prejudice a pet theory, a philosopher is no more to be trusted in such a matter than a thief is to be trusted in a witness-box as evidence in his own favour. Even had this been "sometime a paradox, the time" would "now " give it proof." It is not that the philosopher will lie like the thief, in wilful misreport of his consciousness; but by the very conditions of the case, unless he be one of a thousand, he is incapable of an accurate observation and candid notation of its contents. Unawares, he cannot help looking at his consciousness through the coloured medium of his theory. It is as if by a series of abstruse reasonings a man who had never seen snow should convince himself it must needs be a green substance, and then obstinately refuse to look at it except through a pair of green spectacles. Obviously it would be difficult to convince such a man of his error; the rather as of course he would be capable of solemnly asserting it was merely a calumny of the unconverted to say he wore spectacles at all. And that Mr Mill, who, any time these twenty years, rather piques himself on the success with which he has argued freedom out of the world, is *now* incapable of regarding his consciousness, in so far as it is said to testify of it, except through causational spectacles, is obvious in the very form of his objection: "I am told that if I elect to murder, I am conscious that I could have elected to abstain. *But am I conscious that I could have abstained if my aversion to the crime and my dread of its consequences had been weaker than the temptation?*" A query which is really equivalent to this—Am I then to admit in consciousness the thing which I see, if admitted, to convict me of an unsound process of reasoning? Mr Mill's deliverance is not so properly a denial of freedom in consciousness, as a synopsis of Necessitarian argument. Again—"I therefore dispute altogether that we are *conscious of being able to act in opposition to the strongest present desire or aversion.*" We may reasonably resent this intrusion into the simple question of consciousness, of the writer's theory of action as inevitably in every case determined by a balance of the motives of desire and aversion. The consciousness of freedom asserted is simply our conviction, previous to act, of our being able at will to choose either of two courses of conduct; and subsequent to our choice of one course, of a power in us to have chosen the other. Whether such a consciousness exists is one question, and it seems an eminently simple one; it asks merely a *yea* or a *nay*, on a candid self-interrogation. It is another question, and also a very simple one, whether such a consciousness can in logic give good account of itself. The first is a question for plain men;

the other a question for philosophers; and in either case
there is great unanimity in the answer. Plain men make
no doubt about their freedom of choice; and philosophers (fools excepted) are equally at one as to their entire
inability to explain it; the stoutest asserter of Freedom,
who says proudly with Tennyson, "Our wills are ours,"
adding with him in humility, "*We know not how.*" *
" *How*, therefore," says Hamilton, for instance, "moral
" liberty is possible in man or God we are utterly unable
" speculatively to understand." The questions, kept apart,
are simple; but considerable embarrassment arises when
confounding the distinct points of view of the plain man
and the philosopher, a reasoner like Mr Mill persists in
answering the one question in terms suggested by the other.
When Mr Mill disputes our "consciousness of being able
" to act in opposition to the strongest present desire or
" aversion," he seems to dispute a mode of consciousness
which is scarcely that asserted; inasmuch as neither before nor after act, of the relative strength of our desires
and aversions, is there any clear measure in consciousness. Speculatively, it is evident that the act cannot but
follow the strongest desire; but this merely speculative
point of knowledge is no part of our original consciousness, and is rigorously to be kept distinct from it. The
question in its proper simplicity is this—Have we, or
have we not,—the relative strength of the desires being

* "Our wills are ours, we know not how;
Our wills are ours, to make them thine."—*In Memoriam.*

undetermined in consciousness—a feeling or conviction that of two courses of conduct we are free at will to choose either; and afterward, that having chosen one, we could at will have chosen the other? Assuming, as of course we must, that in any two given philosophers the facts of consciousness are the same, if in a matter so simple as this, one of them answers yes, while the other puts in a negative, it can only be because one of the two palters with his consciousness, whilst professing a candid interrogation of it; and which of the two it is that does so, cannot be considered doubtful if the appeal is to the general consciousness. Let us put to Mr Mill a case, and it may be well to make it one of those "frivolous" cases, in such dubieties recommended by Adam Smith, on the ground that "in them the judgments of mankind "are less apt to be perverted by their systems." The question shall not be whether Mr Mill is to murder or not to murder, but one much less momentous. It is premised that within the space of a minute, let us say, Mr Mill is to put his finger to his nose; the alternative in act proposed to him is whether he will elect to put it to the right side of his nose or the left—the instance is a somewhat too homely one, yet the better on that account. Would Mr Mill, without bringing any of his logical great guns to bear upon it, tell us simply in this case of his *consciousness?* That though the motives are here refined to an almost inappreciable point of subtlety and evanescence, they are present under as utter a rigour of law as if they were obvious and urgent; and further, that Mr Mill, however we suppose him to act, could only have acted otherwise, in virtue of a change in the antecedents, involving a change in prior antecedents, so that, the right side of his nose being touched, the whole previous order of the universe must have been altered to enable him to touch its left side—these are of course obvious deductions from the law of universal Causation, but they have nothing to do with consciousness. Sweeping his mind quite clear of them, if such a thing at all be possible, would Mr Mill report to us of *that* in its simplicity? Is not he conscious of being able to touch at will either the right side of his nose or the left? Having touched, let us say, the left side, is not he conscious he could have touched the right had he so willed it, and conscious that he *could* have so willed, chosen, or *preferred?* If Mr Mill admits himself so conscious, he admits all that is desired of him, for precisely such a consciousness it is, that all unsophisticated men acknowledge, and would think it ridiculous to deny, throughout the whole domain of voluntary action. Should Mr Mill, on the other hand, deny that he is so conscious, we venture to assert with some confidence, that his consciousness contradicts that of every man not a Necessitarian philosopher; and further, that it is not his veritable consciousness, but a fraudulent substitute palmed off upon him by the "system" to which he is wedded. Farther, as there cannot be a doubt we have a consciousness of *some* kind, in regard of our actions, if it be not a consciousness of Freedom, it must needs be a consciousness of Necessity. And will the assertion of such a consciousness find favour with any mortal except perhaps here and there a zealot of the doctrine, only capable of looking at the matter through some mist of his reasoned preconceptions?

But though Mr Mill, in denying for himself all consciousness of being able to act otherwise than as he does act, is at issue with the mass of unsophisticated mankind, he is at one with them in admitting in certain cases, a consciousness that he *ought* to have acted otherwise, and involved in this, a sharp sense of shame and self-condemnation in not having done so. In other words, he admits as a fact of consciousness moral obligation and Responsibility. Let us proceed to inquire whether to the consciousness he admits, we can rationally attribute validity, divorced from the consciousness he denies —whether this *ought* to have acted otherwise, with its accompanying sanctions of shame and self-reproach— can, apart from a *power* to have acted otherwise, be admitted as anything but dream and illusion? whether *blame* can in reason be attributed to acts, the agents being helpless to avoid them? But before entering on this central part of the discussion, it may be well to say something of a passage which Mr Mill, in proceeding to treat of it, interpolates with the label—Important. " Another fact which it is of importance to keep in " view, is, that the highest and strongest sense of the

"worth of goodness and the odiousness of its opposite, is
"perfectly compatible with even the most exaggerated
"form of Fatalism. Suppose that there were two pecu-
"liar breeds of human beings—one of them so constituted
"from the beginning, that however educated or treated,
"nothing could prevent them from always feeling and
"acting so as to be a blessing to all whom they ap-
"proached; another, of such original perversion of nature
"that neither education nor punishment could inspire
"them with a feeling of duty or prevent them from being
"active in evil-doing. Neither of these races of human
"creatures would have free-will; yet the former would
"be honoured as demigods while the latter would be re-
"garded and treated as noxious beasts; not punished,
"perhaps, since punishment would have no effect on
"them, and it might be thought wrong to indulge the
"mere instinct of vengeance; but kept carefully at a
"distance, and killed like other dangerous creatures, when
"there was no other convenient way of being rid of them.
"We thus see that even under the utmost possible exag-
"geration of the doctrine of Necessity, the distinction
"between moral good and evil in conduct would not only
"subsist, but would stand out in a more marked manner
"than now when the good and the wicked, however
"unlike, are still regarded as of one common nature."

Except that from "the highest and strongest *sense*" of moral distinctions there can be no valid inference of their *reality,* inasmuch as the sense *may* be an illusory one, and in the scheme of Necessity *must* be so, as we hope to be able to prove—" these considerations" are certainly, as Mr Mill says, "pertinent to the subject," and would, if acquiesced in, go some way to make further argument on Mr Mill's part unnecessary; but he must positively be asked to *re*-consider them. Mr Mill here either writes something to which we have difficulty in attributing a meaning, or he would not hesitate to speak of an *immoral* toad, tiger, or rattlesnake; in which case, we should not be surprised to hear him, in virtue of his presumed partiality for mutton, proceed to expatiate with enthusiasm on the sublime moral excellence of sheep. Already in the contrasted demeanour of the tiger and of the sheep, the one of which would kill and eat Mr Mill without scruple if only a chance were given it, whilst the other placidly submits to be killed, in order that by Mr Mill it may be eaten, the "distinction between good and evil in con-" duct," as conducive to *benefit* or *injury*, is illustrated in a manner which Mr Mill must hitherto have found not less "marked" than agreeable. But has "the distinc-" tion between *moral* good and evil in conduct" been any way thus illustrated? Scarcely, we should say, unless tigers and sheep are to be elevated to the dignity of *moral agents*. And unless Mr Mill's two supposed "peculiar breeds of human beings," beneficent and malign respectively, are conceived of as moral agents, however their supposed actions might illustrate in a novel

and " more marked manner than now," the " distinc-
" tion between good and evil in conduct," the distinction
between " *moral* good and evil in conduct" would no
more be set forth by them than it now is by the actions
of tigers in contrast with those of sheep. Mr Mill then
must be presumed to consider—if his argument is to be
worth anything—that his contrasted " peculiar breeds "
are moral agents. But how can he possibly do so?
Any one wishing to give an exact definition of a homicidal maniac, could not do better than adopt *verbatim*
Mr Mill's account of his malign " peculiar breed;" and
of course, in admitting that the members of it " would
" be regarded and treated as noxious beasts," he explicitly
abolishes their *moral* agency, except in so far as that of
tigers and rattlesnakes may be moral; the which, if Mr
Mill seriously considers it, he is like to have a monopoly
of his opinion. As to the other " breed," though it suits
Mr Mill to assume that they would be " honoured as de-
" migods," it seems quite as likely, we think, they would
be merely despised as idiots. Acting from a blind, irrational, undiscriminating impulse of benevolence, they
would simply be amiable maniacs; and as such they
would be certain to be regarded; their philanthropy,
like other philanthropies we have heard of, would
speedily be seen to be a nuisance, as disturbing the
moral order, of which it could be no part, and a check
might have to be put upon it. We venture to think it
probable that Mr Mill's " demigods" would be looked

upon as fit only for the mad-house. It is certain that no more than the other "breed," who "would be re-"garded and treated as noxious beasts," could they be looked upon as in any sense moral agents.

How then, on Mr Mill's hypothesis, would "the dis-"tinction between *moral* good and evil in conduct not "only continue to subsist, but stand out in a more marked "manner," &c.? Over the area covered by the hypothesis, the distinction would be plainly obliterated in the obliteration of all moral agency whatever. Outside of that area, of course, the distinction would "continue to "subsist" as before it did; and it might even in this sense be "more strongly marked" that the evil actions of creatures in all else so resembling ourselves would be likely to be specially impressive, on obvious principles of association. Thus it is that the murder of his mother by a maniac would shock us inexpressibly more than if the old lady had been carried off and eaten by a tiger, though *morally* the man would be no more culpable than the brute. So much we may admit, but it makes nothing for Mr Mill. The special thing to be enforced "as against him is, that over the area of his hypothesis moral agency is obliterated. If we desire him, as we surely may, to extend that hypothesis, so as to include the whole human species, what then? A "murder grim and "great," in which the human demons give speedy account of the "demigods," by their nature as defined, incapable of injuring even their enemies, and proceed to "chaw up" each other; the world is at once a huge Bedlam on our hands, and moral agency or agent nowhere in it survives. Mr Mill's argument pushed to its conclusion is thus as neat and complete a specimen of argumentative suicide as perhaps could be readily adduced—unless, indeed, as we said, tigers, toads, and maniacs, are to be dignified as moral agents. "It is of importance to "keep in view that the highest and strongest sense of the "worth of goodness, &c., is perfectly compatible with even "the most exaggerated form of Fatalism!" Who doubts that it is so compatible as regards their own good and evil, in minds presumed outside of the fatal circle? Certainly no advocate of the doctrine of the Freedom of the Will. But surely, if a Fatalistic hypothesis is to be worth anything as illustrating moral distinctions, it must be *within* the fatal round of it that such distinctions must be proved to emerge; and within the round of Mr Mill's hypothesis, they would plainly have no place; for whatever the distinctions between the acts of creatures to whom no moral agency is attributed, *moral* distinctions they could hardly be.

This unfortunate passage is virtually an attempt on Mr Mill's part to turn to account of his argument his favourite doctrine of Utility, as determining the moral qualities of actions; in which light it may perhaps bear to be a little further scrutinised. The Utilitarianism here by implication set forth is really so extravagant a form of the doctrine, that we can scarcely think

Mr Mill would seriously undertake to maintain it against criticism. That Morality and Utility are, within their common area, coincident, is on all hands admitted; and though a whole school of thinkers refuse to admit that *therefore* Utility is the ultimate ground of Morality, Mr Mill can undoubtedly, as against them, make out a very strong and plausible case in favour of that opinion. But surely Mr Mill himself does not so understand the relations between Utility and Morality, as to consider the ideas *co-extensive*. Holding that good moral actions not only subserve utility, but are only good *because* they subserve it, does he also hold that all utilities are moral? Not so, of course; otherwise, to use Adam Smith's illustration, a chest of drawers must be an object of his moral approval, in as strict a sense as Howard the philanthropist. Actions then it is that are moral, and the actions not, we should suppose, of all creatures, but of some, and with full decisiveness the actions of human creatures only. That in certain of the domesticated animals, some emergence of what might seem a moral nature may be noted, is too obvious to be denied; and Coleridge, who could not be happy without some millstone of a mystery to peer into, says somewhere he finds in the dog a deep one, in virtue of that dawning of a human conscience which he plainly apprehends in the brute. But in our necessary ignorance of brute psychology, which we can only, as across a gulf, guess at from external signs,—the analogical interpretation of these being plainly of the most precarious character,—we must for ever be at a loss to determine how far what seems a germ of morality in the brute, is really so in such a sense as fulfils the human definition, and how far it may be merely of the nature of an automatic simulation of it. Wherefore, to the actions of men only can we decisively attribute moral qualities, and these we are supposed to discriminate as good or evil, approvable or condemnable, according—not necessarily, as we *see* them to conduce,—a misapprehension of the utilitarian doctrine which vitiates very much of the argument directed against it even by intelligent critics,—but as they may belong to the classes, which on a wide induction *have been found* to conduce to human welfare or its opposite. But in strictness, it is of agents that we morally approve or disapprove, not of actions; or only of actions as they seem to illustrate a moral quality in the agents. Moral acts presuppose moral agents; to say that there can be moral acts without moral agents, is only a shade less glaringly absurd, and not any whit less really so, than to say that acts can take place without agents. Thus it is, that to the act of a maniac, bearing precisely the same relation to utility as that of a sane man, we attribute no moral quality whatever. To constitute an act moral, it must—apart from its tendency to subserve utility or the reverse—be done morally; that is, in fulfilment or outrage of a known law of duty or obligation;* and as the maniac is amenable to no law of any kind save that of causational necessity, we absolve both him and his act from all stigma of moral blame. Mr Mill, however, would distinguish between him and his act, absolving the agent, yet classifying his act as *immoral*. Obviously so; for as we said, the homicidal maniac corresponds most exactly in definition with the

---

* That there is nothing here which conflicts with Mr Mill's Utilitarianism, though something perhaps which does with those coarser statements of the doctrine which Mr Mill implicitly repudiates, may be seen in these snatches from his little volume on the subject :—" So far as to " external sanctions. The internal sanction of duty, whatever our stan- " dard of duty may be, is one and the same,—a feeling in our own mind; " a pain more or less intense attendant on violation of duty, which, in " properly constituted moral natures, rises in the more serious cases into " shrinking from it as an impossibility. This feeling, when *disinterested*, " and connecting itself with the pure idea of duty, is the essence of Con- " science." Again,—"The ultimate sanction of all morality (external " motives apart) being a subjective feeling in our minds, I see nothing " embarrassing to those whose standard is utility, in the question, what is " the sanction of that particular standard? We may answer—the same " as of all other moral standards—the conscientious feelings of mankind." Elsewhere—in his Essay on Bentham—we find him sharply noting as a defect, "the absence of recognition, in any of his writings, of the exist- " ence of Conscience as a thing distinct from philanthropy, from affection " for God or man, and from self-interest in this world or the next."

Whether in these and similar passages Mr Mill does not implicitly sub-insert for his Doctrine of Utility the ultimate moral basis, its need of which is denied, may perhaps admit of question. He must be held to admit in them, self-interest wholly apart—into some remoter form of which the straitest sect of Utilitarians make no scruple of reducing our feeling of duty—an obligation of duty or Conscience, a *moral* obligation, to promote utility. Were we to ask Mr Mill *whence* this obligation?

imaginary "breed of human beings," by nature maleficent, whose supposed actions would, according to Mr Mill, illustrate in a very marked manner "the distinc"tion between *moral* good and evil in conduct." This arbitrary disjunction of the agent and his act, as if the one could be talked of as moral without inference of morality in the other, Brown, in his Lectures, with almost weary iteration, comments upon, as more than

we confess we are curious to know how precisely he would answer. To allege Utility *itself* as the ground of the obligation to pursue it would be surely to run round like a mill-horse in rather a vicious circle. It seems to us we must therefore assume as basis of the doctrine a natural bond between man and man, leading us to pursue Utility defined as the happiness of individuals and of the race. That the root of the matter is here was seen sharply by Hume, who accordingly smuggles in a note to this effect,—"It is needless to push our researches so far as "to ask why we have humanity or a fellow-feeling with others? It is "sufficient that this is experienced to be a principle in human nature. "We must stop somewhere in our examination," &c. Hume, like others, stops where it is convenient for him to stop; but the further inquiry seems pertinent, whether this primary principle in human nature be not in human nature self-guaranteed as a *moral* principle? If it be so, as perhaps there are fair grounds for maintaining, Morality is a primitive fact of human nature, and no mere growth of experience, however in its after development experience may modify its details. The fundamental objection to Utilitarianism, or the Doctrine of Social Expediencies, is that it *assumes* Society, which, except as a product of primitive moral forces, might have found it not easy to constitute itself. In the very idea of Society, the idea of Morality is involved; in the social affections and impulses in man, there is furnished the moral germ, out of which his subsequent Moralities are developed. But people are wont in these times to speak and write loosely of "development," as if no living germ were needed for it; as if in the notion of a growth we did not pre-suppose something to grow.

anything else having tended to confuse the whole theory of Morals.* To this disjunction, however, Mr Mill stands fully committed not only in this passage by inference, but distinctly as thus, in his work on Utilitarianism,—" Utilitarian moralists have gone beyond all "others" (and that Mr Mill goes along with them to the full is clear from the context) "in affirming that the "*motive* has nothing to do with the morality of the "action, though much with the worth of the agent." Wherefore, as by the "worth of the agent" Mr Mill cannot possibly mean anything but his *moral* worth, he holds that a moral act can take place without a moral agent, and is directly at issue with Brown. He is also at issue with a greater than Brown, to wit, David Hume. He it was who first, so far as we know, reduced to scientific system the scattered facts, which, taken together,

* "An action, though we often speak of it abstractly, is not, and "cannot be, anything which exists independently of the agent. What "the agent is, as an object of our approbation or disapprobation, that his "action is; for his action is himself acting." "Much of the confusion "has arisen from the abuse of one very simple abstraction,—that by "which we consider an action as stripped of circumstances peculiar to an "individual agent, and forming, as it were, something of itself, which "could be an object of moral regard, independently of the agent." "It "is no small progress in Ethics, as in Physics, to have learned to distin- "guish accurately abstractions from realities, to know that an action is "only another name for an agent in certain circumstances"—and passages to the like effect occur every few pages. Of course in the mere matter of statement, objection may be taken to the *dictum* that "an "action is only another name for an agent;" but the essential meaning of Brown here is not thereby impugned.

approve the moral Doctrine of Utility; and as an exposition of the subject in detail, the treatise in which he did so has not since perhaps been quite equalled. His opinion, as that of an advocate of Mr Mill's doctrine, is plainly, in this particular, of somewhat more weight as against Mr Mill, than that of Brown, its express opponent; and thus distinctly it is given:—"*Actions* are "*objects of our moral sentiment,* so far *only* as they are "indications of the internal character, passions, and affec- "tions; it is impossible that they can give rise either to "praise or blame where they proceed not from these prin- "ciples." And this is certainly the view of the matter which has hitherto commended itself to the common sense of mankind. That it will yet a while continue to do so seems on the whole likely. If it be really necessary that, in order to believe in the Doctrine of Utility, we must first believe that motives have nothing to do with the morality of actions, and that moral acts can take place without moral agents, however clearly it may approve itself to the minds of a few philosophic *illuminati*, the mass of men, till the day of doom, will have none of it. That this disjunction of the morality of the agent and that of his act is essential to Utilitarianism, as taught at least by Mr Mill, is by no means clear to us; but to his present argument it *is* essential. The moral qualities of agents, it is implied, are no measure of those of actions, which must be estimated as moral simply as they subserve utility; and the action retains its moral quality, even when the agent is an irresponsible maniac. But having gone so far, can we avoid going a little further? Having extended our conception of Morality so as to include the acts of homicidal maniacs—in definition identical with one of Mr Mill's "peculiar breeds," from whose supposed actions he illustrates Morality anew— we cannot logically stop there; we must needs go on— as Mr Mill seems in this passage to make no scruple of doing—to include in it the act of the man-eating tiger, of the *cobra* which stings the man who treads on it, of the toad which creeps into our arbour and squats there, of the bug which disturbs our rest o' nights when we go into a London lodging. To any one who should gravely propound such a scheme to us, we might feel inclined to reply by holding forth with corresponding gravity on the morality of a pair of breeches. Inasmuch as the breeches, though *useful*, are certainly not *active*, as the bug most uncomfortably *is*, our irony would be unphilosophic, we know, in its neglect of an important distinction, but we almost think we should venture it.

The following from Mr Bain's "Emotions and the "Will," before glanced at, may be quoted as having here a certain pertinence—see page 564. "Every animal "that pursues an end, following up one object, and avoid- "ing another, comes under the designation of moral. "The tiger chasing and devouring his prey, any creature "that lives by selecting its food, is a *moral agent*." Mr Bain seems here directly at issue with our argument pursued above, and in harmony with that of Mr Mill; but we cannot suppose him to be really so. He is remarking on the ambiguity of the word " moral," as in " Moral Philosophy," which " is on one supposition "confined to Ethics, or Duty, and on the other compre-" hends, if not the whole of the human mind, at least the " whole of the Emotions and Active Powers;" and in relation to this ambiguity he goes on to say—" In the " large sense I am a moral agent when I act at the insti-" gation of my own feelings, pleasurable or painful, and " the contrary when I am overpowered by force. It is " the distinction between mind and the forces of the " physical world, such as gravity, heat, magnetism, &c.; " and also between the voluntary and involuntary activities " of the animal system. It would be well if the same word " were not indiscriminately applied to two significations " of such different compass; for there can be little doubt " that perplexity and confusion of idea have been main-" tained thereby. Still nothing can be better established " than the recognition of both significations, and we are " bound to note the circumstance that the ' moral ' which " at one time coincides with the ' ethical,' at other times " is co-extensive with the ' voluntary.'" That some "per-" plexity and confusion of idea" exists here on Mr Bain's part is evident on a comparison of the last sentence with his previous one. " Every act that follows upon the " prompting of a painful or pleasurable state, or the asso-" ciations of one or other, is a voluntary act, and is "all that is meant, or *can be meant* by moral agency." Yet just before we have seen that "the 'moral' at one "time coincides with the 'ethical'"—wherefore a voluntary act is *not* "all that *can* be meant by moral agency," which term *may* be used to indicate acts which have in addition to their voluntary quality an "ethical" one; and in truth cannot with any accuracy be used otherwise. Except as in "Moral Philosophy," and such phrases as "moral causation," "moral certainty," &c., the "moral" is distinguished from the "physical," Mr Bain has here, if we mistake not, invented the confusion he complains of. We are aware of no authority for his use of the word "moral" as meaning simply "volun-"tary;" nor would any weight of authority justify a writer in continuing to so use it. In voluntary agent—voluntary action—our meaning is sufficiently expressed; why run the risk of embarrassing it by intruding as a synonym the word "moral," already in a twofold sense appropriated? To speak of the tiger as a "moral agent" on the ground that it "selects its food" voluntarily; of the bug which also "selects its food," as some of us know to our discomfort, as also on that ground a "moral "agent," is a hideous abuse of language, which Mr Bain ought not to have suggested. But, allowing him this use of the word "moral" as meaning merely "voluntary," he would probably admit that when the "moral distinctions "of actions" are in question, we must needs mean actions distinguished from merely voluntary ones by the addition of an ethical quality of right or wrong as it may be, and so much admitted, our argument does not any way suffer from his whim of speaking of a tiger as in that restricted sense a "moral agent." The writer must express his regret that in the preparation of his little Essay, he had not the advantage of being acquainted with Mr Bain's very interesting and valuable treatise, which has only come under his notice as these sheets are in course of being printed.

And now of Moral Responsibility, and this latest attempt of Mr Mill to harmonise it, as an admitted *datum* of Consciousness, with his scheme of Causationism, as exclusive of human Freedom, defined an Ability in man to act otherwise than as he does act. "What," asks Mr Mill, "is meant by Moral Responsibility? " Responsibility means Punishment. When we are said " to have the feeling of being morally responsible for our " actions, the idea of being punished for them is upper- " most in the speaker's mind." When Mr Mill, having asked in the first sentence—"what is meant by moral " responsibility?" answers in the second—" Responsi- " bility means punishment;" and goes on in the third to speak of our " feeling of being morally responsible," nobody could at first sight interpret him otherwise than as speaking of Moral Responsibility throughout. But in the light of the distinction he makes a little farther on between " the belief that we shall be *made* account- " able, and the belief that we *ought* so to be," the last of which only it is which " can be deemed to require or presuppose the Free-will hypothesis," there seems ground to suspect that by Responsibility, as used in the second sentence, he means Responsibility *simple,* as distinguished from the Moral Responsibility indicated in the first and third.* But surely, if this be so, Mr Mill in this all important point of definition, is guilty of no little slovenliness. For, of course, he must very well know—as himself frequently so using the word—that when without qualification, we speak of Responsibility, it is Moral Responsibility we expect to be understood to mean. Wherefore, from this in the first sentence he was clearly bound to formally distinguish it in the second, and again to distinguish in the third by " when we are " said on *the other hand,*" or some such equivalent of contrast. The truth might almost seem to be, that indicating the distinction between the *simple* Responsibility and the *moral* one, he was content with a most confused indication of it, as feeling in some semi-conscious way that it would not in the least suit him to

---

* This of course *must* be Mr Mill's meaning. " What is meant by " Moral Responsibility? Responsibility means Punishment. When we " are said to have the feeling of being *morally* responsible for our actions, " the *idea of being punished* for them is *uppermost* in the speaker's mind." But the idea of punishment being *uppermost,* implies of course some correlative idea as *undermost* in the mind of said speaker. Being " *morally* responsible " includes therefore something *more* than is given in the Responsibility which merely "means punishment." Let this distinction by Mr Mill himself—however loosely indicated—be carefully kept in view.

elaborate it into clearness. As the distinction, so simple as it seems, is really one of quite cardinal importance for the argument, let us do for it in a cursory way what it did not suit Mr Mill to do. Any Imperative which has force to make itself effective will constitute a Responsibility in the subject of it; only a Moral Imperative can constitute a Moral Responsibility. To illustrate in an easy way—in the case of a tyrannical schoolboy, who says to a little fellow *not his fag*—for over *him* there would be some semblance at least of right—" Go to my " bed, you young hound! and warm it for me, till I come, " or if you don't, I'll thrash the life out of you,"—an Imperative is announced by the big fellow, and it constitutes a Responsibility in the little one, inasmuch as he knows, if he refuses, he will be made to answer for his conduct. This, as a mere appeal to fear, is the simple or *brute* imperative as distinguished from the human or Moral Imperative, whose appeal is to our moral judgments of merit and demerit in conduct. If, instead of telling him to warm his bed for him, the big fellow had said to the little—" Go rob me Farmer Hodge's orchard, " and bring me hither the apples," the little fellow would have found himself the subject of the Brute and the Moral Imperatives in conflict, and would have had to elect which of them to obey. Again, there are *mixed* Imperatives, in which the two elements may coinhere in proportions indefinitely variable. Such are the authority exercised by a parent over a child, and that of the Law prohibitive of crime on penalty. With the first of these we are not here directly concerned. As to the Law, in so far as its appeal is to fear *simpliciter*, it is important to note it as merely a form of the lower or Brute Imperative; and though in most minds this more or less connects itself with some form of the higher or Moral one, the essential distinction between them is not thereby affected. A caitiff who refrains from murder solely because he dreads being hanged for it, is the mere slave of the Brute imperative; he, on the other hand, to whom "Thou shalt not kill" comes as a clear mandate of his moral nature, is indeed, in virtue of the accompanying threat, also a subject of the lower Imperative, but more or less he is made free of its jurisdiction by the life within him of the higher one. Now, the clear distinction between these, which is untouched by the fact that they may efficiently coexist in proportions varying in various minds, Mr Mill has seen fit to indicate only to confound it. In his preliminary definitions it is slurred; and throughout the subsequent discussion it is never steadily kept in view, the word *responsibility*, (or *accountability*, as it may be,) being used indifferently to imply the *simple* or the *moral* one, so that which of the two is really meant it is sometimes not easy to know. And in thus playing fast and loose with the distinction, Mr Mill has in one sense done wisely, inasmuch as he could not have clearly and consistently exhibited it without exhibiting along with it the inconsequence of his whole argument. For, standing pledged as he does to constitute for us on the principle of Necessity the Moral Imperative, in order to maintain the validity of which it is that his opponents postulate Freedom, he constitutes for us in that argument only the simple or Brute Imperative, our social right to exercise which is inferred from the efficacy of fear *simpliciter* as a motive. This we proceed to show.

"It is not," writes Mr Mill, "the belief that we shall "be *made* accountable for our actions, which can be "deemed to require or presuppose the Free-will hypothe-"sis; it is the belief that we *ought* so to be; *that we are* "*justly accountable; that guilt deserves punishment*. It "is here that the main issue is joined between the two "opinions." And again—"The real question is one of "*Justice;* the legitimacy of retribution or punishment." Mr Mill, having thus stated the "main issue" and the "real question," it seemed reasonable to suppose that accepting these ideas (of "guilt *deserving* punishment"— of *Justice* involving "the legitimacy of retribution,") as given in our moral consciousness, he proposed to show that, in their own nature, said ideas did *not* of needs presuppose the hypothesis of Freedom, but that an adequate basis could be found for them in the rival scheme of Necessity. This, however, is only in part the case. In the course of the discussion, Mr Mill says that he "can find no "argument to justify" punishment inflicted on the principle of "a natural affinity between the ideas of guilt and "punishment, which makes it intrinsically fitting that "wherever there has been guilt, pain should be inflicted "by way of retribution." The "main issue" being as to a "*justly* accountable" in the sense that "guilt *deserves* "punishment,"—"the real question" one of "Justice— "the legitimacy of retribution," it might seem that the idea of Retribution being thus explicitly discarded, that of Justice, by Mr Mill himself given as its moral equivalent, must needs be dissipated along with it, so that there is no longer before us any question whatever. But with Justice Mr Mill is indisposed to part company. Reserving formal remark on this disjunction in the process of his argument of ideas in its premises identified, let us see how Mr Mill succeeds in his attempt to substantiate the idea of Justice.

"The real question is one of Justice,—the legitimacy "of retribution or punishment. On the theory of Neces- "sity, we are told, man cannot help acting as he does; "and it cannot be *just* that he should be punished for "what he cannot help. Not if the expectation of "punishment enables him to help it, and is the only "means by which he can be enabled to help it? To say "that he cannot help it, is true or false according to the "qualification with which the assertion is accompanied. "Supposing him to be of a vicious disposition, he can- "not help doing the criminal act, if he is allowed to "believe that he will be able to commit it unpunished. "If, on the contrary, the impression is strong on his "mind that a heavy punishment will follow, he can,

"and in most cases does, help it. There are two ends
"which, on the Necessitarian theory, are sufficient to
"*justify* punishment; the benefit of the offender him-
"self, and the protection of others."

All which is most lucid and exact, but with "the
"real question of Justice," whereof Mr Mill supposes
himself to be discoursing, has literally nothing whatever
to do. The Justice of which Mr Mill stands pledged
to treat is Moral Justice, in its severe regard of the
past, inflicting punishment as *deserved;* and for this
Mr Mill, as he proceeds, quietly substitutes a simple
expediency as respects the future.* The feat of logical legerdemain is facilitated by an ambiguity in terms.
There is Justice as above defined; there is the justice
of a remark or an anticipation; we speak of a just
(as a fit or expedient) arrangement; and we consider
we justify an action when we prove it negatively permissible. Any one may see at a glance that in the
passage quoted, Mr Mill's reasoning succeeds in at all

* We, of course, cannot be unaware that in this reference to expediency we may be held to be running our head inadvertently against Mr
Mill's Utilitarian theory of Morals; but that Mr Mill in his own person
cannot object to the distinction made here, will be evident from these
snatches from his work on the subject—"This seems the real turning-
"point between morality and simple expediency"—"The distinction be-
"tween the feeling of right and wrong and that of ordinary expediency,"
—and more explicitly—"Is then the difference between the *Just* and
"the *Expedient* a merely imaginary distinction ? By no means. The ex-
"position we have given of the nature and origin of the sentiment recog-
"nises a real distinction."

connecting itself with "the real question" solely by a disgraceful jumble of such various meanings of the word. Mr Mill, when asked whether it be *just*, *i.e.*, *deserved*,* in the sense in which "guilt deserves punish-"ment"—that a man should be punished for what he could not help in the past, considers he replies in the affirmative, when he says it is expedient to punish him, as we flog an uncleanly cur to improve its manners in the future. Of the following instance of the confusion what is to be said?—" Free-will or no free-will, it is *just* " to punish men for this purpose, (of protection,) exactly " as it is *just* to put a wild beast to death for the same " purpose." This only, that these same wild beasts to which Mr Mill seems partial, have been known in the rage of hunger to make away with a philosopher; and farther, that the man who should speak of a shot tiger, as having "*justly* met its doom," *exactly* in the sense of the words as applied by him to a human miscreant, would simply prove he had either some twist in his moral perceptions, or was grossly ignorant of language. Again—what of this? "To punish a man for his own " good, provided the inflicter has any proper title to con-" stitute himself a judge, is *no more unjust than to ad-*

* "It is universally considered *just* that each person should obtain "that (whether good or evil) which he *deserves;* and unjust that he "should obtain a good, or be made to undergo an evil which he does "not deserve. This is perhaps the clearest and most emphatic form in "which the idea of Justice is conceived by the general mind."—*Vide Mill's* "*Utilitarianism,*" page 65.

"*minister medicine!*" Has the giving of medicine to a patient any relation whatever to *Justice?* Clearly not—unless, indeed, the patient has paid a fee for it, in which case the doctor would doubtless be *unjust,* if he fraudulently refused to fulfil his part of the bargain. Otherwise, however it might be proper and humane to administer medicine, improper and inhuman not to administer it, to speak of its being given or withheld as either *just* or *unjust,* is merely to pass an outrage on the accepted meaning of terms. It is easy for a man at this rate allowed him, to prove whatever he pleases. If, in using the word white, he means in one sentence white, blue perhaps in the next, and green as it may chance in a third, naturally he puzzles you a little, somewhat as Mrs Quickly did Falstaff—you "know not " where to have him." "In the present case," writes Mr Mill a little previously, "there is more than a verbal " fallacy, but verbal fallacies also contribute their part." Undoubtedly they do; verbal fallacies contribute, in point of fact, the main part of Mr Mill's argument.

"Not if the *expectation* of punishment enables him " to help it," &c. This clause has, indeed, a correspondence with " the belief that we shall be *made* accountable " for our actions," by Mr Mill himself, as we saw, set aside as irrelevant to the discussion; its relation to the " main issue" involved in our " being *justly* made " accountable," and the fact " that guilt *deserves* punish- " ment" is by no means quite so obvious. Of the simple or Brute Imperative, justifiably (permissibly) announced by society for its own ends, an instance is here admitted; but of the Moral Imperative there is as yet no hint; and the only Justice involved is a consideration of social expediency, which justifies (or makes permissible) our announcement of the Imperative of fear; a consideration in no respect of principle to be distinguished from that which is the ground of our flogging a cur to teach it to respect the carpet. The cur alike and the man here are mere slaves of the Brute Imperative. "Sup-"pose him to be of vicious disposition," and so on. In nothing of all this have we got beyond the first rude motive of fear, as we may find it urgent on occasion in the basest of the lower animals. Of the higher, or Moral Imperative, the keenest partisan of Mr Mill will seek in vain for a trace throughout his entire discussion, saving as he seems at times to imply that out of the primitive element of the Brute Imperative, it is possible psychologically to *grow* the Moral one; a point which will afterwards turn up for consideration.

Further—"There are two ends which, on the Neces-"sitarian theory, are sufficient to justify punishment; " the benefit of the offender himself, and the protection " of others." That it may be *justifiable* to punish a man for his own good, as we conceive it, even though he should not *deserve* punishment, is perhaps with limitations to be admitted—limitations which we should expect Mr Mill, of all men, as the author of the noble

Tractate "On Liberty," to recognise very sharply. But solely in virtue of the man's *deserving* the punishment can we accurately call it *just*. To justify any course of conduct, punitive or other, is, as we have before said, to prove it negatively permissible—to prove it not in any flagrant sense *unjust*; it may also, of course, be to prove it positively proper, its propriety having been called in question. But countless things are proper, and even obligatory, the obligation to which is not one of Justice; countless things are not *unjust*, to speak of which as *just* could be only to incur ridicule. But wherefore labour a matter, Mr Mill's fallacy in regard of which is obvious in his own illustration, already in the casual way glanced at? "To punish him for his own good is *no more unjust than* "*to administer medicine.*" To administer medicine to a patient is not *unjust*; also it is not *unjust* to shave one's-self of a morning. Yet—the question of fee in the former case apart—the relation to positive Justice is, in the two cases, identical; it is in both a purely negative relation—no relation at all. Here, as throughout the discussion, it is the whim of Mr Mill to disport himself some miles from "the real question," as we have seen it by himself indicated. Again, as to punishment which has for its end "the protection of others," it seems plain that, justly or even justifiably—*pace* Mr Mill, there is between the two a slight distinction—we could not inflict it on a man who had in no sense *deserved* punishment. On occasion of a murder, for instance, the real culprit being *non inventus*, we could scarcely clutch hold of a rough at random and hang him for it. For *protective* purposes—the public presumed ignorant of the little deceit practised on it—the innocent rough would be excellent; yet Justice for those behind the scenes would surely receive in his dying struggles no very sublime illustration. Clearly, as it seems to us, for no social end to be served by it, could we justly inflict punishment, save in virtue of that evil *desert* in the subject of it, which constitutes the punishment in some sort a retributive one. Touching this evil desert it is, that, according to Mr Mill, " the main issue is joined between " the two opinions." But nowhere does Mr Mill join issue on it; he evades it here and throughout. Of " punishment as a precaution taken by Society in self- " defence" here is accurately the sum of what Mr Mill has to say. "To make this just, the only condition " required is, that the end which society is attempting " to enforce by punishment should be a just one. Used " as a means of aggression by society on the just rights " of the individual, punishment is unjust. Used to pro- " tect the just rights of others against unjust aggression " by the offender, it is just. If it is possible to have just " rights, it cannot be unjust to defend them." Readers must be left to find their way as they can in this jungle of justs and unjusts; we must really decline to enter it with them. Yet, timidly treading its outskirts, let us just glance at the last sentence. It is possible, we suppose, to have "just rights;" but a master of English like Mr Mill would, we think, be unlikely to use such a phrase, unless under the mask of a tautology he wished to insinuate a sophism. Did ever any one hear of unjust rights, or of just wrongs? Farther,—the rights being admitted just rights, not unjust ones, we are told " it cannot be unjust to defend them." And of course it cannot, if justice be consulted in the means employed to defend them. But it is possible to defend them *unjustly*. As instance already given, we might defend them most effectively by hanging innocent people, or poor unhappy maniacs. But to such a display of social " vigour" Mr Mill would almost certainly decline to give the sanction of his great name. Consequently, what he must be held here to mean is, not that our " just rights"—such of our rights, that is, as do not chance to be wrongs—" it cannot be unjust to defend " by any means that may lie to our hand; but that it cannot be *unjust* in us to defend them *justly*. The tautology in the first clause of the sentence is thus, it seems, of no avail for Mr Mill's purpose, except in virtue of a suppressed tautology in the second, too flagrant to be put upon paper. Surely further remark is here unnecessary; the more so, that, as in the previous case, Mr Mill's own concluding illustration, already *en passant* commented on, sufficiently defines his position. " Free-will or no free-will, it is just to punish for " protection *exactly* as it is just to put a wild beast to

" death for the same object." In this identification, in so far forth, of the man and the wild beast, there is a ground, of course, for the announcement of the Brute Imperative; but we had thought it was the Moral Imperative which Mr Mill was engaged in elucidating for us.

When we said above that the consideration of *desert* was throughout evaded by Mr Mill, we were not rigorously accurate. It is touched lightly in *one* passage, which must be quoted, otherwise we should be liable to the suspicion of wilful unfairness to Mr Mill.

" Now the primitive consciousness we are said to
" have that we are accountable for our actions, and that,
" if we violate the rule of right, we shall *deserve* punish-
" ment, I contend is nothing else than our knowledge
" that punishment will be *just;* that by such conduct
" we shall place ourselves in the position, in which our
" fellow-creatures, or the Deity, or both, will naturally,
" and may *justly*, inflict punishment upon us."

Now, what does Mr Mill mean by *Justice, Just,* &c.? We don't ask what in the course of his discussion he means—for this is a metaphysical inquiry of quite too subtle and complex a kind—but what is he *bound* to mean by the words as by himself defined in his premises? " The real question is one of Justice; the legitimacy of
" Retribution." " The belief that we are *justly* account-
" able; that guilt *deserves* punishment. It is here that
" the main issue is joined between the two opinions."

Just punishment is certainly here *deserved* punishment. Also in the note at page 52 we have elsewhere seen Mr Mill shut himself up to this meaning of the word; and if another note to the like effect be needed for behoof of desert and Retribution, the reader has it below.*

Wherefore, if anywhere in his discussion Mr Mill is found using the word *just*, except as including the idea of *desert*, unless in doing so he defines the precise shade of difference in the new meaning intended, he is to be held utterly inexcusable, as condescending to use for our bewilderment the stalest trick of the sophist. Holding Mr Mill, then, strictly to his own definitions of the word *Just*, what in effect do we here find him writing? "The " consciousness we are said to have that we shall *deserve* " punishment, I *contend*, is nothing more than our " knowledge that punishment will be *deserved;* that our " fellow-creatures, &c., may *deservedly* inflict punish- " ment on us." No great need to *contend* here; the most contentious opponent Mr Mill is likely to encounter will scarcely care to dispute with him as to a deserved punishment being a deserved one. And now, does or does not Mr Mill in his argument maintain that

---

* " Retribution, or evil for evil, becomes closely connected with the " sentiment of Justice, and is universally included in the idea," (p. 89.) " The principle, therefore, of giving to each what they *deserve*, that is, " good for good, evil for evil, is not only included within the idea of " Justice, as we have defined it, but is a proper object of that intensity " of sentiment which places the Just, in human estimation, above the " simply Expedient," (p. 90.)—*Mr Mill's* " *Utilitarianism.*"

"guilt deserves punishment?" We profess that after reading Mr Mill repeatedly with our best care, we do not in the least know; and it is even our notion, it might puzzle himself very much to say whether he does or not. The following dilemma is proposed to him, in terms of his own initial statement of the dispute,—If guilt does *not* deserve punishment, what becomes of his argument for Moral accountability, as distinguished from the accountability simple, which merely "means punish-" ment," or the being "*made* accountable?" Plainly it is nowhere; in the "issue joined between the two " opinions," the one of which holds that, Freedom apart, Moral Responsibility is without a basis; the other, that without Freedom, it is still logically to be substantiated, —the latter, or Mr Mill's opinion, has experienced ignominious rout, and no longer holds the field. On the other hand, if a man thoroughly does *deserve* punishment, might he not in *equity* be punished, simply and finally on that ground, apart from those ulterior ends— " the benefit of the offender himself, and the protection " of others"—by which alone punishment, according to Mr Mill, can "on the Necessitarian theory be justified?" And if not, in what intelligible sense can he be said to *deserve* punishment? Such are the hopeless difficulties in which—unless language is to be emptied of meaning in his favour—Mr Mill has here contrived to enmesh himself.

Tedious as it is to track Mr Mill through all his

doublings, and oppressive as the reader may be beginning to find this part of the subject, it seems necessary to proceed with it a little. To quote Mr Mill further,—" If any one thinks there is justice in the infliction of
" purposeless suffering; that there is a natural affinity
" between the two ideas of guilt and punishment, which
" makes it intrinsically fitting that wherever there has
" been guilt, pain should be inflicted by way of retribu-
" tion; I confess that I can find no argument to justify
" punishment inflicted on this principle. As a legiti-
" mate satisfaction to feelings of indignation and resent-
" ment, which are on the whole salutary, and worthy of
" cultivation, I can in certain cases admit it; but here it
" is still a means to an end. The merely retributive
" view of punishment derives no justification from the
" doctrine I support."

"The *merely* retributive view of punishment." But does not in this word *merely*—if it be taken to mean anything—Mr Mill virtually concede a retributive *element* at least in punishment; an evil *desert* in the culprit, failing of which, not even as "a means to an end" could we justly visit him with punishment? Is not something of the kind implied too in describing as " legitimate" the " satisfaction to feelings of indigna-
" tion" conveyed in the infliction of punishment? and could the satisfaction be held " legitimate," such punishment being supposed *undeserved?* Unless by his guilt a man *deserves* punishment, could we justly for our own

ends inflict it? And if *guilt* be admitted to deserve punishment, is not this equivalent to an admission of the "natural affinity between the ideas of guilt and "punishment," above by Mr Mill denied? At the same time, while we contend that the idea of punishment as in any accurate sense *just*, is that of punishment deserved, and as such, in some strict sense, retributive, "*merely* retributive punishment" we should perhaps not venture to inflict, except in the indulgence of personal passion, or strong sympathy with such in others. Otherwise, no more than Mr Mill can we "find argument to "justify it." Even when clearly convinced of its *justice*, we might yet shrink from it as *unjustifiable*, on the general principle of humanity, which forbids " the in- " fliction of purposeless suffering." Mr Mill's point of view here is that of every Christian Moralist. The " vengeance is mine, I will repay," peremptorily forbids the infliction of punishment, even in equity due, when there is not a concurrence of social ends to be served by it, with that evil desert of the offender, which constitutes the punishment a just one.

Finally, Mr Mill thus rids himself of the subject of Justice, in a passage which, whatever his previous success in it may be, he plainly holds to clinch his argument, (p. 512.) " I ask any one who thinks that the " justice of punishment is not sufficiently vindicated by " its being for the protection of just rights, how he recon- " ciles his sense of justice to the punishment of crimes

"committed in obedience to a perverted conscience?
"Ravaillac and Balthasar Gerard did not regard them-
"selves as criminals but as heroic martyrs. If they were
"justly put to death, the justice of punishment has no-
"thing to do with the state of mind of the offender,
"further than as this may affect the efficacy of punish-
"ment as a means to its end. If that is not a justifica-
"tion, there is no justification. All other imaginary
"justifications break down in their application to this
"particular case."

Apart from Mr Mill's habitual confusion of the different shades of meaning of the word *Just*, and its derivatives, Justice, Justify, &c., there seems nothing here which to an antagonist of Mr Mill needs present the smallest difficulty. As to "crimes committed in obe-
"dience to a perverted conscience," it seems sufficient to say that we consider them justly (or deservedly) punished *as* so committed; we hold the felon responsible for his crime, if not immediately perhaps, yet mediately as culpable in the perversion of his conscience which led to it, in so far as this may fairly be surmised to have emerged under the conditions of sanity. Of assassins who "regard themselves not as criminals but as heroic
"martyrs," we may boldly say that could we have positive assurance that their outrage of the obligation to respect life was solely an act of self-sacrifice to what they considered a higher and more sacred one, however, on obvious grounds of general expediency, we might

acquiesce in the doom awarded them, the Justice of it as *deserved* or *due* to their deed, considered in itself, and as an isolated act, we should very peremptorily deny. *Justifiable* we should call it in the general, not *just* in the particular instance. Take the stock case of Brutus —his purity of motive unimpeached—who, inasmuch as he did not love Cæsar less, but Rome more, struck through his own heart at the murderer of his country's freedom—of Charlotte Corday, who smote the monster Marat! To deeds like these—except that "the God-"like stroke" of Brutus seems a little perhaps *too* God-like—humanity throbs for evermore approval. What of such cases? Was Charlotte Corday "*justly* put to "death," in the sense that she *deserved* her doom? Mr Mill will not say so. *Justifiably*, however, from the legal point of view, she was, if to the ruffianly faction then dominant it be proper to ascribe legality. Generally, in such cases, while we may doubt if it be morally *just* (deserved) that the particular hero should suffer for what may really have been an act of sublime virtue, his punishment may yet seem *justifiable* to us, on the ground that such whimsical heroes are inconvenient, inasmuch as no society could afford to grow a succession of them. A dubious point of Justice—dubious, because the true motive of the act must always remain obscure—may here be allowed to be over-ridden by a plain and potent mandate of expediency. Further, a judicial and punitive *system* can only deal on general rules laid down, with overt act, and can take no account whatever of the motives and " state of mind of the offender." But, has *Justice*, therefore, as Mr Mill alleges, " nothing " to do with the state of mind of the offender?" That it is not so, is obvious from the *one* exception to these general rules, which in every humane judicial system is made in favour of the maniac. *His* " state of mind " is most carefully taken into account; and if it be not on a ground of Justice that it is so, on what ground is it? Mr Mill will find it hard to explain. His saving clause, "further than as this (his state of mind) may " affect the efficacy of punishment as a means to its " end," is here of course of no avail. For how should the state of mind of the maniac, as unamenable to motive, any way affect the efficacy of our hanging him for murder, as a means to deter others from murder? Not surely—for anything it will make in Mr Mill's favour—by directing the sympathies of the public *against* the law and in favour of its victim—for what is to be assigned as the ground of this supposed public sympathy and abhorrence? Plainly nothing but a sense of the brutal *injustice* of hanging a man, " the state of whose " mind" fatally determined him to crime. Yet unquestionably, rotating on the rope as an example, he would be as edifying as the most perfect sanity could make him; and were expediency only considered, to be rid of the pernicious confusions in which we are of late involved, we should begin to hang madmen to-morrow. But the sense of *Justice* to the poor wretches, as not after all *deserving* to be hanged, which at first procured their exemption from the rope, will probably be strong enough to perpetuate it, even against a very strong plea of social utility as outraged. That " the Justice of Punishment " has nothing to do with the state of mind of the " offender," is a *dictum* which Mr Mill may profitably revise, in the light of this significant exception. If in other cases we take no account of " the state of mind of " the offender," it is because we are utterly incapable of doing so with any approach to scientific accuracy, and because criminal legislation *can* only proceed on a general rule of particular penalties attached to particular acts. But not unfrequently it may happen that to a punishment which is *justified* under the general rule, we can with difficulty reconcile ourselves as *just* in the individual instance. And this, because though legally we cannot, unavoidably as men we *do*, regard the " state of mind " of the offender " as determining the character of his crime, and more or less affecting our estimate of the amount of punishment *due* to it, or in the special case, *just*.

But all this time has Mr Mill been so juggling with Justice without some dim sense of what he is doing? Not so; in the midst of his dexterous legerdemain, a sudden suspicion seems to emerge in his mind that all is not quite as it should be, and his attempt to escape from the entanglement in which he finds himself, is, in the mode of it, even more lamentable than was the entanglement itself. After reducing, as we have seen, "the primitive consciousness we are said to have that we are accountable, and that if we violate the rule of right we shall *deserve* punishment," to "nothing else than our knowledge that punishment will be *just*," by which he really means no more than that it will "naturally follow;" though "will naturally and may *justly* follow" are of course the terms used, a little qualm seems to strike him, and he says,—" By using the word *justly*, I am not assuming in the explanation the thing I profess to explain. *As before observed*, I am entitled to postulate *the reality*, and the knowledge and feeling of moral distinctions. *These*, it is both evident metaphysically, and notorious historically, are *independent of any theory concerning the will.*" Had we, in this connexion, met with the above on the page of an unknown writer, we should have held it to indicate the most radical philosophic incompetence. Philosophers would have easy times of it if we allowed them quietly to postulate the very things they undertake to prove. If Mr Mill's present chapter be not in its main scope an attempt, on the hypothesis of Necessity, to *prove* "the *reality* of moral distinctions," as involved in human Responsibility, failing of which man cannot be a moral creature, of whose acts such distinctions are predicable, we will, with any sauce or none, *eat* his entire discussion. That on the scheme of unconditional Necessity, as including human actions, moral qualities can no longer, except fictitiously, be attributed to them; and that thus " moral distinctions," however practically accepted, must needs be illusory, not *real*—what if not this is the objection urged by the advocates of Freedom against Mr Mill's doctrine? Could Mr Mill himself state the objection, save implicitly in these terms? Except by ingeniously *mis*-stating it, he could not; yet in answer to the objection, he considers himself " entitled to pos-" tulate the *reality* of moral distinctions." We venture to question his title to do so. "As I have before " observed, I am entitled," he says; but even in this he is mistaken; nothing that he has before observed amounts to any claim to be so entitled. The reference in his "before observed" can only be to two passages. The one is that important one already quoted, in which, seeking to make sun-clear to us the "distinction between " *moral* good and evil in conduct," he considered he succeeded in doing so by abolishing all moral agency; the other is a passage in which he says, that for the present argument, it is no matter what " criterion of moral " distinctions," utilitarian or otherwise, is assumed. In the first of these, he asserts that " the highest and strong-" est *sense* of the worth of goodness, and the odiousness " of its opposite, is compatible with even the most exag-" gerated form of Fatalism;" in the other he says, " it " is sufficient if we *believe* there is a difference between " right and wrong." The " feeling (sense) of moral "distinctions," and "the knowledge" of them—if he chooses to call it a knowledge, rushing into those confusions of knowledge and belief of which he elsewhere convicts Sir W. Hamilton—he has thus, in his previous observations, entitled himself to postulate. And *these* he is welcome to postulate; but the *reality* of moral distinctions is quite another affair. That, in the scheme of the Necessitarian, moral feelings and beliefs—since nobody questions their existence—may logically be assumed as involved in the all-including Necessity, we are frankly ready to admit; we shall further admit as involved in it the conviction in the mind of Mr Mill that between a belief and the validity of a belief, as vouching the *reality* of the thing believed, there is no distinction whatever; but this last part of the Necessity seems for Mr Mill an unhappy one. We had thought a mere belief could prove nothing beyond *itself*; that even if it were one of Sir W. Hamilton's "natural beliefs," it was scarce even therein proved Natural. Now, it seems, *nous avons changé tout cela*, and beliefs are to be put in the witness-box as evidence to their own character. Whether does Mr Mill, from the *belief* in moral distinctions, find so sure an inference of their *reality* as seems to entitle him to postulate it, or has he of that reality an absolute intellectual intuition? We should be curious indeed to be informed. It seems to us that to clutch at it in the latter way would be quite as consonant with his general scheme of thought as to try to arrive at it in the other. In one sense, however—let him reach it in what way he pleases —Mr Mill does wisely and well to postulate the *reality* of moral distinctions, as without something of this surreptitiously implied in the words just, justly, &c., when in truth their meaning as used excludes it, he is really without an argument. In another sense, he cannot be held to do quite so well, inasmuch as the trick is quite too transparent; so that, for any but the dullest reader, his dexterous manipulation of terms is vain as the art of the thimble-rigger when the dupes have discovered his mystery. As to the "evident metaphysically that " the *reality* of moral distinctions is independent of " any theory concerning the will," what is to *us* metaphysically evident in the matter, is that precisely *this* it is, which Mr Mill is engaged in trying to prove, not, as we think, with highly-distinguished success. Suppose, again, it *were* " notorious historically"—as to a certain extent it is—that the "knowledge (belief) and feeling of " moral distinctions are independent of any theory," what is this to Mr Mill's purpose? It is also, we suppose, "notorious historically" that the extreme speculative idealist, who regards the external world as an illusion merely, does not therefore endanger his supersubtle brain by plunging his head against walls. He practically believes in the wall as having a most serious external existence, though speculatively he admits it as nothing but an idea of his own absurd mind. And as against Mr Mill it is contended that in the scheme of Necessity, consistently developed, our moral belief is precisely in analogous case; however practically urgent, it must speculatively be volatilised into a lying illusion of life; the "moral Imperative" can only be recognised as a mendacious one, but not the less it may remain an *efficient* Imperative.

When Mr Mill, having defined "the real question as "one of Justice—the legitimacy of Retribution *or* Punishment," proceeds from his premiss of Necessity—with what success we have seen—to substantiate the idea of Justice by proving the legitimacy of Punishment, either he has stated the question in terms of most serious inaccuracy, or he must be held to be proving, along with this, the legitimacy of Retribution; for if in his *formula* —"Retribution *or* Punishment"—the words be not used as interchangeable, it can only be because Mr Mill is an indifferent writer of English. But the sole Punishment his reasoning legitimates, is Punishment reformatory and deterrent; and of Retributive Punishment we hear nothing except towards the close of this part of the discussion, when Mr Mill in an easy way informs us he "can find no argument to justify it." Not the less, however, in terms of his own statement, he stood bound to find argument to justify it; and till we came with some surprise on his disclaimer, we confess we had thought he was trying to do so, though he did not seem much to succeed. The disclaimer, when it comes, is well; but meantime, for careless readers—who are greatly, we fear, in the majority—Mr Mill's unquestionable success in legitimating punishment, interpolated as motive, has *seemed* to involve its legitimacy when inflicted as also Retribution: and even when the disclaimer is tendered, they mayn't, perhaps, appreciate its full force as confounding the whole previous argument. We would not for the world hint such a thing of Mr Mill; but in a writer presumed capable of it, we might almost have been disposed to suspect here a little dishonesty of artifice. Moreover, as we before noted, had Mr Mill applied to the subject, as *defined in his own words*, the least of his usual rigour, " Retribution," being thrown overboard, " Justice" alike and " Punishment" should have been cast out along with it, and instantly " the real question" would have been no longer question at all to tax Mr Mill's ingenuity.

Of artifice or subterfuge of any kind we hold Mr Mill incapable; but throughout he has been led by the exigencies of an attempt, in its very nature now and for ever a hopeless one, into something perilously like it. Recurring to his definition of Moral Responsibility— " when we are said to have the feeling of being morally " responsible for our actions, the *idea of being punished* " for them is *uppermost* in the speaker's mind"—it is obvious to remark, as before noted, that " the idea of " punishment being *uppermost*" implies some correlative idea as *undermost* in the mind of said speaker. Yet, in the discussion which follows, this *undermost*, plainly implied, finds no recognition whatever. The *uppermost* exclusively has place in it, and the " Responsibility means " punishment," which defines the *simple* Responsibility, is made to do duty throughout as if it had defined the *moral* one. In this way Mr Mill's argument succeeds beautifully. " Responsibility means punishment;"— and Mr Mill, finding it easy to exhibit a pretty little *schema* of thought, justifying, on the principle of necessitated action by motive, the infliction of punishment *as* a motive, of course he proves his case; Necessity and Responsibility are no longer incompatible, as they were held to be; Mr Mill has prospered to a wish in the mixing of his oil and water. But if, on the other hand, " Responsibility means Punishment," which suggests the simple Imperative, is inadequate to suggest the Moral one, it may very well be that though the fluids have been by Mr Mill most vigorously shaken together, their interfusion is semblant only, and may presently be proved to be so, by the obstinate re-emergence of the oil. And precisely this is the case. If Moral Responsibility means *not* Punishment *simpliciter*, but Punishment in some sense *retributive*; and if, with "the idea of being pun- " ished *uppermost* "—since Mr Mill will have it so— beneath and bound up with this idea is the sense of being *justly* punished, of *deserving* to be punished, Mr Mill's whole argument collapses. And that such is the accurate statement of the case—that our feelings of good desert in a good deed, of evil desert in an evil one, are of the intimate essence of Moral Responsibility, no sane creature will deny, save here and there, perhaps, a philosopher. Mr Mill himself will scarce deny it; for did not we find him remark that "*not* the belief that we shall be *made* accountable, but the belief that we *ought* so to be, that we are *justly* accountable, that *guilt deserves punishment*, can be *deemed to require or presuppose the Free-will* hypothesis?" This, therefore, is the Responsibility, by Mr Mill himself given as the fact of consciousness, to furnish a rational basis of which his adversaries "deem it needful to presuppose Freedom;" and this, by consequence, it plainly was, which the rigour of the debate required Mr Mill to constitute for us, under his prescribed conditions of Necessity. His utter failure to do so, to the extent of having in the end to admit—after trying to bubble us with a discourse on Punishment Reformatory and Deterrent, quite futile and beside the "real question"—that "for any natural affinity between the two ideas of guilt and punishment he can find no argument," is perhaps the severest misfortune which ever befell a logician of Mr Mill's admitted eminence.

Sensible, it might almost seem, of some little awkwardness in the admission to which he is led,—that the retributive view of punishment "derives no justification from the doctrine I support,"—Mr Mill, by way of effecting a diversion and carrying the war into the camp of the enemy, adds immediately—" But it derives quite as little from the Free-will doctrine. Suppose it true

"that the will of a malefactor when he committed an
"offence was free, or *in other words*, that he acted badly
"not because he was of a bad disposition, *but for no
"reason in particular*, it is not easy to deduce from this
"the conclusion that it is just to punish him. That *his
"acts were beyond the command of motives* might be a
"good reason for keeping out of his way or placing him
"under bodily restraint; but no reason for inflicting
"pain upon him when that pain, by supposition, could
"not operate as a deterring motive." Surely these words
in italics convey a serious misrepresentation of the Free-
will doctrine, as stated by any intelligent supporter of
it; nor, unless Mr Mill can produce from some such admittedly competent supporter of it a *dictum* that to act
in freedom, and to act without a motive, and "for no rea-
"son in particular,"—that is, from the mere wild irrational
caprice of a maniac,—are terms in meaning identical, will
it be easy for us to acquit him of seeking a cheap victory
with his readers, by the easy method of substituting for
the opinion to be discredited a questionable figment of
his own. No advocate of Freedom has ever yet said
that men act badly not because they are of bad disposition, but *for no reason in particular;* idiocy of this kind
is too palpable to be promulgated except as the wisdom
of an opponent. Under stress of a hopeless argument
which sought to make liberty *comprehensible*, foolish
things may have been written; Reid, in particular, is, in
nearly all his remarks on the subject of motive, incom-

petent; and when—contending, as he does, that it is possible for the will to act without a motive—he asks casually in this relation—" Is there no such thing as "wilfulness, *caprice*, or obstinacy among mankind?" he really seems, in some helpless way, to glance in the direction of Mr Mill's "for no reason in particular;" but his lapse will scarcely avail for Mr Mill, who, in citing these words of Reid, appends to them with express approval, Sir W. Hamilton's brief and contemptuous dismissal of them. After this, it needed some coolness on Mr Mill's part, in an argument originating as against Hamilton, to interpolate as accepted Free-will doctrine, a somewhat audacious caricature of Reid's rejected suggestion of caprice. Moreover, not only is Mr Mill's definition of freedom in itself objectionable, as conveyed in irrational terms, and attributing to his opponents an opinion which to a man they would repudiate as fatuous,* but the use he proceeds to make of it lies open to still more serious objection. It might almost seem that this passage, which identifies freedom of action with action undetermined by motive, and done "for no reason in

---

* Many men will say that at any given time they feel themselves free to eat or not to eat; but will any man say that when he does eat, he eats not because he is hungry, but "for no reason in particular?" Plainly there is no such blockhead. In the cases again in which hunger is not the reason of his eating, a man knows he has some other reason, failing which he would not eat. *How*, if this reason is sufficient to make him eat, he should yet have been free not to eat, no competent person professes his ability to explain.

" particular," was mainly written by Mr Mill to give
occasion for the Note he appends to it, as follows :—
" Several of Sir W. Hamilton's admissions are strong
" arguments against the alleged self-evident connexion
" between Free-will and accountability. We have found
" him affirming that a volition not determined by mo-
" tives 'would, if conceived, be conceived as morally
" 'worthless;' that 'the free acts of an indifferent are
" ' morally and rationally as worthless as the pre-ordained
" ' passions of a determined will;' and that ' it is impos-
" ' sible to see how a cause undetermined by any motive
" ' can be a rational, moral, and accountable cause.' If
" all this be so, there can be no intuitive perception of a
" necessary connexion between free-will and morality:
" it would appear, on the contrary, that we are naturally
" unable to recognise an act as moral, if it is, *in the sense
" of the theory*, free."

In the whole range of philosophical controversy, we
should be surprised if there could be found an instance
of greater unfairness than this strange passage involves.
In *the sense of the theory* free! *What* theory, may we
ask, and *whose?* The theory, of course, of motiveless
volition and action "for no reason in particular." And
whose theory of Freedom is this? Hamilton's, one would
naturally suppose, if, as against Hamilton, any weight,
or indeed meaning, is to be attributed to Mr Mill's remarks. But how stands the matter? Plainly, in these
passages quoted, Hamilton, who has previously (p. 498)

been exhibited, as showing, that if the will be acted on by motives, "we can never *in thought* escape determina-"tion and Necessity," is now, on the other hand, enforcing, that a will which should act *without motives*, would, "if conceived," be conceived of as something other than the reasonable and moral will, with which only his argument concerns itself as worth contending for as a Will at all. Free-will, before proved inconceivable on the hypothesis of motive, is now, on the negation of that hypothesis, shown to be conceivable—if at all—as merely itself a negation. Hamilton is here, in fact, according to his special whim of elaborating the argument, negatively presenting Freedom as one of his two famous opposed Inconceivables, in terms nearly identical with those used by Leibnitz in his curious controversy with Dr Clarke,—"A mere Will, without any "Motive, is a *Fiction;* not only contrary to God's per-"fections, but also, *chimerical and contradictory; incon-"sistent with the Definition of the Will,* and sufficiently "confuted in my Theodicæa." Similarly Hamilton,— "Nay, were we even to think as true, what we *cannot* "*think as possible,* still the doctrine of a *motiveless voli-*"*tion* would be only casualism; and the free acts of an "indifferent," &c., as by Mr Mill quoted. The theory of free volition, as motiveless, and "for no reason in "particular," is thus plainly enough not Hamilton's, who throughout admits, and even proclaims, that of Freedom, human or divine, no positive logical *theory* can be substantiated,—"*How*, therefore, I repeat, moral Liberty in God or man is possible, we are utterly unable speculatively to understand." *Whose*, then, is this theory of free volition, as "beyond the command of motives, and for no reason in particular?" It seems to originate with Mr Mill, who, finding his antagonist without a theory of Freedom, is kind enough to furnish him with one *gratis*. Yet in a manner he takes payment for it; confidently parading in the text this paltry figment of his own, as the approved doctrine of Freedom, and by plain implication, Hamilton's, he cites, in his note, as "admissions" by Hamilton the very passages in which that figment is shattered, and turns these so-called admissions against his genuine doctrine of "the necessary connexion between Free-will and Morality," by reviving the demolished figment. "It would appear," he says, (from these admissions,) "that we are naturally unable to recognise an act as moral, *if* it is in the sense of the *theory*, free;" but it is the very purpose of said admissions to explode the theory, as unthinkable, contradictory, and absurd; wherefore, Mr Mill's important *if* having no virtue whatever in it, his whole Note collapses wretchedly, leaving "the necessary connexion between Free-will and Morality" precisely as it was before. Mr Mill, having otherwise not much satisfactory to say against this "necessary connexion," and perhaps half uneasily feeling as much, the surreptitious character of his attack upon it here, is to be looked at

a little leniently. It is the more strange to find Mr Mill here writing as he does, that a little before we have found him writing—" The inconceivability of the Free-
" will doctrine is maintained by our author (Hamilton)
" not only on the general ground just stated, (of its in-
" volving an absolute commencement,) but on the
" further and *special* ground *that the will is determined*
" *by motives.*" After this to attribute to Hamilton, the *theory* that the will is " motiveless " and acts " for
" no reason in particular " seems a little to stretch the licence of controversy. Doubtless, it is held by Hamilton that Free-will, though inconceivable, as involving the impossible notion of " an absolute commencement " —of a cause which is not itself an effect—is yet, on his principle of the Excluded-Middle, to be *believed*. But it is only *as* an Inconceivable he claims our belief in it, not as an impossible "absolute commencement," as Mr Mill seems to suppose;* and out of the abyss of the

---

* How, while with emphasis asserting the determination of the will by motives, could Hamilton also intend to assert "an absolute com-
"mencement" as the *mode* under which Freedom, though "inconceiv-
"able," was yet to be believed? This would have been to rush with his eyes open on the staring contradictory of a thing at once caused and uncaused. That Hamilton has not done this is evident from his own words candidly interpreted—" Were we *even* to think as *true* (to believe)
" what we cannot think as *possible,*" which seem fairly to imply, that to insist on our belief in Freedom under this impossible or unthinkable mode is no part of his doctrine. Free-will he shows to be inconceivable, first, on the motiveless hypothesis, as involving an unthinkable "absolute commencement;" and again, on the special ground of the

Inconceivable to which he consigned it,—had he further elaborated the rough outline of the argument, which is all that we have from his hand—he would no doubt have sought to educe its possible reconcilement with the plain and undeniable fact, by himself, as we saw, insisted on, that "the will is determined by motives." To represent him in any case, as denying this, and by implication maintaining that men act badly, not because they are of bad disposition, but "for no reason in par-"ticular," is to impute to him gratuitous absurdity. But to impute gratuitous absurdities to an opponent—more especially if he happens to be a dead opponent—is sometimes the safest way of trying to confute him.

fact, by no sane man to be denied, "that the will is determined by "motives." Plainly, it is under this last known and admitted mode, not under the other impossible one, that Free-will must be believed, though "inconceivable;" acts are determined by *motives*, (caused) and yet are to be believed in as in some utterly "inconceivable" manner *free*. That this is really the doctrine which Hamilton, in the last resort, would have held himself bound to maintain, is made plain by the following passage:—"This philosophy brings us back from the aberr-"tions of modern theology to the truth and simplicity of the more "ancient Church. It is here shown" (proposed to be shown) "to be as "irrational as irreligious, on *the ground of human understanding*, to "deny either, on the one hand, the foreknowledge, predestination and "free grace of God, or on the other hand, the Free-will of man; that "we should believe both, and both in unison, though unable to com-"prehend either, even apart." Substituting here for the theological terms Foreknowledge and Predestination, their philosophical equivalents of Causational sequence or Necessity, we find that, taking this with Freedom, we must "believe *both*, and both in unison," on grounds *extra*-logical. Yet these, oddly enough, are our old friends the

It must be held curious, that, whilst giving us an elaborate theory, from the point of view which suits his case, of Punishment as an external institution, Mr Mill has ignored almost utterly throughout what it specially behoved him to explain—that primary type of punishment which is given in the inner phenomena of conscience. When we find Mr Mill writing as we have seen—" When we are said to have the feeling of being " morally responsible for our actions, the idea of being "punished for them is *uppermost* in the speaker's mind," it cannot but occur to us, that if by " punished" he means externally punished, he lays a somewhat undue emphasis on what is really, in terms of his own statement, the inessential side of the phenomenon, " the belief that we shall " be *made* accountable," which nobody " can deem to " require or presuppose the Free-will hypothesis." Doubt-

---

two opposed Inconceivables, to belief in *one* of which we found ourselves before shut up by the logical "Law of the Excluded Middle." We are inclined, as before hinted, to agree with Mr Mill in thinking that Hamilton's reconstruction of the argument in the light of his " Philosophy of the Conditioned," has really done it no great service; but it was not right to represent him as, in order to maintain freedom, denying motives and causes. In the light of the rest of his remarks, the one casual expression from which there might seem an inference of his doing so, " the inconceivability of an absolute commencement, on " the *fact* of which commencement the doctrine of liberty *proceeds*," must be held a mere inadvertancy of expression. By fact here Hamilton means, and *can* only mean, a notional fact, as distinct from a fact of existence; a notion which, even in trying to " think it as *true*," we " cannot think as *possible*," yet the only notion of freedom competent to us when we seek to evolve it in the *process* of thought.

less there are human hounds, in whose minds the "idea of
"being punished" is not only uppermost as a motive deterring them from crime, but is even the sole deterring motive; but probably when even a very average specimen of the race speaks of being morally Responsible for his actions, it is not "the idea of being punished for them" that is uppermost in his mind, but the idea of doing wrong, and incurring, to adopt Mr Mill's words elsewhere quoted, "the pain more or less intense attendant on the "violation of duty." Punished or not for sin as *externally* we may be, *within* we cannot escape punishment; the shame which haunts us for a conscious meanness, the remorse which dogs us for a crime—from *these* we should vainly seek to flee, though from every other form of punishment certain of absolution. This "primary type" of punishment, the intimate sense of *desert* in act, Mr Mill would perhaps not seek to deny; neither with the smallest hope of success could he attempt to deny it *retributive*. That whilst retributive, it is also motive, is at once frankly admitted; but how does it succeed in *becoming* motive except by its first experienced efficiency as retribution? We should be curious to know how Mr Mill would deal with this primary type; and yet, strictly, we are *not* very curious, inasmuch as, from glimpses here given, we can accurately predict how he would. Not being able to deny it as a fact of consciousness, he would deny it as a *primitive* fact, and undertake implicitly to reduce it for us by ex-

hibiting it as a synthesis of previous understood elements.

Mr Mill has not at all seen fit to labour this part of his subject; and the few hints he has dropped in relation to it, may be very briefly disposed of. "It is well "worth consideration," he says, (p. 508,) "whether the "practical expectation of being thus called to account "has not a great deal to do with the internal feeling of "being accountable; a feeling assuredly which is seldom "found in any strength in the absence of that practical "expectation." Of this it seems enough to say that, admitting not only that the expectation of being called to account has *everything* to do with our feeling of being *accountable*, as in truth the very ground of the feeling, but that it has "a great deal to do" with our sense of being *morally* accountable, this does not amount to an identification of the *simple* imperative and the *moral* one. These, as we before explained, may efficiently co-exist, and as intervolved in the same mind, have a very "great deal to do" with each other; the higher imperative, in itself perhaps apt to be sluggish, being urged into healthful activity by a sting of suggestion from time to time applied by the lower one; but not the less they must be held to be principles in their nature distinct, till it is shown that the higher can be exhibited as a mere derivative and development of the lower. This Mr Mill with perfect assurance asserts it can, but he is shy of trying to prove it. "No one," he

writes, " who understands the power of the principle
" of association, can doubt its sufficiency to create out
" of these elements (of anticipated punishment) the
" whole of the feeling of which we are conscious. To
" rebut this view of the case would require positive evi-
" dence; as, for example, if it could be proved that the
" feeling of accountability precedes, in the order of de-
" velopment, all experience of punishment. No such
" evidence has been produced or is producible. Owing
" to the limited accessibility to observation of the mental
" processes of infancy, direct proof can as little be pro-
" duced on the other side; but if there be any validity
" in Sir W. Hamilton's Law of Parcimony, we ought
" not to assume any mental phenomenon as an ultimate
" fact which can be accounted for by other known pro-
" perties of our mental nature." Certainly we ought
not; but *can* the moral idea be thus accounted for? and
does Mr Mill imagine his dogmatic " no one who under-
" stands can doubt" is to be received in settlement of
such a question? *We* presume to *doubt* very much, at
the risk of being held not to *understand;* and our pre-
sumption may seem less in the matter than otherwise it
might have done, when it is found that all Mr Mill has
to say towards dispelling doubt is as afterwards, (p. 513.)
" From our earliest childhood the ideas of doing *wrong*
" and of punishment are presented to our mind together,
" and the intense character of the impressions causes
" the associations between them to attain the highest

"degree of closeness and intimacy. The only ideas pre-
"sented have been those of *wrong* and punishment, and
"an inseparable association has been created between
"these directly without the help of any intervening
"idea. This is quite enough to make the spontaneous
"feelings of mankind regard punishment and a *wrong-
"doer* as naturally fitted to each other,"—a passage
amusing in the effrontery with which a writer engaged
in the *genesis* of the moral idea *assumes* it in the very
phraseology he employs. In his use of the word *wrong*
here with its moral associations, Mr Mill deliberately begs
the whole question. Let us inquire what is really given
in the child's experience. There is given a particular
course of conduct—punishment as its invariable result;
and the mere *word* wrong, which by degrees it appropriates to that course of conduct, just as the word *black*
it appropriates to denote black objects. But no more
than the word *black*, would the word *wrong*, as denot-
ing said course of conduct connote for the child a moral
attribute; all it could connote for it would be the punishment, invariably annexed. And association, were it
thrice over "inseparable," could not generate from the
elements of experience a *moral* element which in experience it did not *find*. What association would no
doubt be able to generate would be an instinctive and
inevitable *expectation* of punishment, as in the very
nature of things attached to the particular conduct;
but the sense of *desert* in the conduct, and of punish-

ment as *justly* attached to it, might still perhaps remain to be accounted for. That Mr Mill can account for such a sense, on his principle of experience and inseparable association, may or may not be; what is certain in the matter is, that here he has not done so, except for unwary readers, by assuming it in the terms employed: and in this way it would be easy to account for almost anything on almost any principle. This feeling of good or evil desert, Mr Mill would not perhaps explain by saying it is *taught us*, (though there are hints here which look in that unhappy direction;) for, supposing he did so,—not to mention that the possibility of teaching implies in the mind taught what Coleridge would have called a " preconformation " to the idea communicated, —the *quis ipsos custodes* would instantly suggest as pertinent the query—Who taught the teacher? how did the idea *originate?* But in truth, for our present purpose, it is not worth a pin's head of pains to inquire whether or no Mr Mill could succeed with his *genesis* of the Moral idea; for supposing him to have succeeded, what then? He would only have accounted for Morality *as an illusion*, by showing us how it became so; he would simply have constructed the illusion for us. And, in doing this, Mr Mill would scarce perhaps be capable of considering he had proved it to be a Reality; though as to this there may be some doubt, in the light of certain of his previous remarks, (*vide* pp. 67–69.) That, on the hypothesis of Necessity, Morality *cannot* be proved so, and can only be speculatively admitted as at best a necessary illusion, a very little trouble will make abundantly obvious.

In his attempt, on the principle of Necessity, to legitimate the Justice of punishment, we found Mr Mill writing as follows:—" Supposing him to be of a vicious "disposition, he cannot help doing the criminal act if "he is allowed to believe that he will be able to commit "it unpunished. If, on the contrary, the impression "is strong in his mind that a heavy punishment will "follow, he can, and *in most cases* does, help it." It is not a little significant that here Mr Mill, in putting two cases which are beside the real inquiry, omits to consider a third, which is obvious, pressing, and pertinent to it. "In *most cases does* help it." Let us take one of the very numerous remaining cases, in which, despite the punishment known to be imminent, he does *not* help it. The question in this case occurs—*Could* he have helped it, or could he *not?* That he *could* have helped it, the antecedents being supposed constant, Mr Mill will scarcely allege, as to do so would be to admit—Freedom. Consequently, Mr Mill must needs elect the alternative that, do what he would, (an unmeaning phrase,) he could *not* have helped it. The farther question then emerges—Can we, in Moral Justice, proceed to *hang* him, let us say, for the crime he could in no sense help? Mr Mill's reply that, on the whole, suppose a man murders his grandfather, it will be proper to tie him up for it, to protect his grandmother who survives, is an audacious shirking of this question, to which common sense and humanity alike (as in the case of maniacs before glanced at) give unhesitating answer in the negative. If the man could not help his action (as Mr Mill must needs admit, or accept the alternative of Freedom,) could rigorously *not* help it, any more than a stone unattached can help falling to the earth, (and will any man with a head more capable than that of a pin pretend, in the all-including, unconditional Necessity he announces, modifications, and degrees of stringency?) all the philosophers who ever bothered the *pia mater* of perplexed mankind might be frankly challenged to produce a ground of moral blame against the man, which would not, by parity of reason, suffice to convict the stone if by chance it had brained a navvy. With precisely as much *justice* (of the *sense* of the thing we say nothing) might you arrest the stone, put it in the dock, try it, condemn it, and finally, carting it to the scene of its misdemeanour, *hang* it as a lesson to the rest of the quarry to respect the brains of navvies, as so proceed against the man. Solely by postulating in the man some quality *not* in the stone, which transcends the necessity common to both,* can we conceive of his act

---

* Observe, no intelligent advocate of Freedom feels called upon, in making his assertion of it, therein to deny Necessity. He believes *both*, each on its appropriate evidence; and in his admitted inability to *reconcile* them in his belief, he sees only one more proof of what already

as crime, and not simply as Fate or misfortune. And what transcends Necessity must be Freedom. We should scarce have thought it possible that any man worth one's while to reply to would propound as the difference between man and stone which legitimates moral judgment of the one and not of the other, the inherence of a *will* in the man; if *ex hypothesi*, that will must *itself* be held to act

> "Not willingly, but tangled in the fold
> Of dire Necessity, whose law"

determines its minutest decision. But we find to our grief and consternation, in Mr John Stuart Mill, such a man. From a passage already incidentally glanced at, there seems little doubt that, in the last resort, it is thus Mr Mill would seek to escape from the difficulty. At p. 514 we find him thus writing:—" Yes; if he really 'could not help' acting as he " did; that is, if his *will* could not have helped it; if " he was under physical constraint, or under the action

---

is proved an hundredfold—the limitation of his logical faculty. That it was to some such issue as this that Hamilton would have conducted his argument, had he consistently—or perhaps inconsistently—developed it—Mr Mill will no doubt prefer the latter phrase,—there is evidence in the passage at page 81 quoted. As it is, all that we have from him on the subject is a quantity of *disjecta membra*—hints, notes, fragments, written at different times, on which it is difficult to ground a conclusive criticism. Curiously enough, Hamilton—fragmentary always—is nowhere more fragmentary than on this topic of Freedom, to which he attributed such importance.

"of such a violent motive, that no fear of punishment
"could have any effect; which, if capable of being as-
"certained, is a just ground of exemption, and is the
"reason why, by the laws of most countries, people are
"not punished for what they were compelled to do by
"immediate danger of death."* A passage this which
seems for Mr Mill an unhappy one. For how, on his
principles, is it possible to maintain a valid distinction
between the exceptional cases given as incapacitating the
will, and thus claiming of right "exemption" from
moral judgments, and other cases for which no such
plea of exemption is urged? In a case of physical constraint, the *will* of the man is in abeyance to the pres-
sure of a physical causation; in a case of such over-

* It is but fair to quote Mr Mill's clause introductory of this passage:—"That a person holding what is called the Necessitarian
"doctrine, should on that account *feel* that it would be unjust to
"punish him for his wrong actions, seems to me the veriest of chimeras.
"Yes, if he really," &c. Questionless, if said person should reason
out the subject as loosely and irresolutely as Mr Mill does, and while
maintaining that he was utterly without *power* to act otherwise than
as he did act, any more than may reside in a stone to help falling
to the ground, should also continue to maintain that he was yet culpable in not having done so, of course it *would* be a chimera. But
otherwise it might not really be so much so as it seems to Mr Mill.
Touching this and related topics, something may be said in the sequel.
Meantime, it is enough for our argument to show, that if we suppose
the man to reason rigorously, and to have rid himself of what he could
only regard as the moral superstitions in which he aforetime was educated, he could not logically look upon himself as *guilty* in his action,
and so a subject of punishment accurately to be called just or deserved:
and this is above sufficiently shown.

mastering motive as no fear of punishment can countervail, the will of the man is in abeyance to that of a *moral* causation; and in either case, Mr Mill holds that, inasmuch as "his *will* could not have helped it," exemption from blame must be accorded. What then of a case in which the motive, though somewhat less absolutely tyrannous, was yet of violence sufficient to determine the man to crime? Is there here no rigour of moral causation? And if in this case, not less than in the other, the causal necessity is admitted, on what ground is the right of the man denied to the "exemp-" tion" granted his fellow? *Could* his will in *this* case " have helped it?"—in *any* case, in which the motive was the sufficient reason of the act? Mr Mill must here be supposed to imply as much, though he could only explicitly maintain it by a plunge into fatal inconsequence. For in *no* case could "the *will* of the man " have helped" his act, except by being determined differently; and if it could not *determine itself* differently, how its different determination could be come at except through that "difference in the antecedents" which Mr Mill himself has taught us to exclude, he will find it hard to explain. How, farther, are we to distinguish between a case in which "no fear of punish-" ment *could* have any effect," and one in which the fear of punishment *had* no effect? If it *had* no effect, how *could* it have had any? Solely by being a *stronger* fear, in relation to the antagonist impulse, to suppose it which, is once more to suppose a change in the antecedents, with a revolution, as involved in this, of the whole previous order of the world. Mr Mill seems here in his argument somewhat in the hapless case of the philosophical gentleman in the *Tempest,* " the latter end " of whose commonwealth forgets the beginning." Briefly, if Mr Mill, to the exclusion of Freedom, includes human actions under the law of Univeral Causation, we are at least entitled to insist on his steadily conceiving it *Universal.* If Physical Causation incapacitates the Will, must not Moral Causation incapacitate? and if not, what is the rational ground of the distinction? Farther, if in special cases, as Mr Mill admits, Moral Causation incapacitates, must it not incapacitate in *all?* and if not, how not? Freedom apart, *could* in *any* case the " *will* of a " man have helped" his doing as he did, any more than his falling to the earth, if he found himself flung forth of window? Mr Mill cannot say so except in manifest outrage of his own principles. The formula of "his " *will* could not have helped it," which he exclusively announces as ground of "exemption" from moral judgments, in cases of physical constraint and tyrannous extremity of motive, must needs, on these principles, be extended to all outlying human actions, with the like inference of "exemption." Mr Mill's " yes—if he really " ' could not help' acting as he did; that is, if his *will* " could not have helped it; if he was under physical con- " straint," on the extension of its second clause, which his own previous reasoning to a change of antecedents necessitates, amounts in point of fact then to this—that in any given case, the man " really could not help " acting as he did, any more than if he had been under physical constraint; how then, any more than in that case, he is to be held a fit subject of *blame,* we may ask Mr Mill to demonstrate, and give him his own time to it. When he has succeeded in doing so, we shall admit his title to be found writing as follows:—

"If the desire of right and aversion to wrong have
" yielded to a small temptation, we judge them to be
" weak, and our disapprobation is strong. If the temp-
" tation to which they have yielded is so great that even
" strong feelings of virtue might have succumbed to it,
" our moral reprobation is less intense. If, again, the
" moral desires and aversions have prevailed, but not
" over a very strong force, we hold that the action was
" good, but that there was little merit in it; and our
" estimate of the merit rises in exact proportion to the
" greatness of the obstacle which the moral feeling
" proved strong to overcome."

Meantime, it is sufficiently clear that the phrases *merit* or *demerit, moral* approbation or reprobation, can except as, so to speak, *stolen,* have no place in Mr Mill's vocabulary. For how should a desire or aversion as failing in the hour of temptation, incur his moral censure as *weak,* if, being, as it is, the last link in a chain of unconditional sequences, we can only suppose it *stronger*, by supposing a change in the series of these sequences? To alter the whole world from the beginning is surely the sort of feat, for his culpable neglect to perform which it seems odd to arraign a poor sinner. Further, in the matter of temptations yielded to, or resisted, why should he apportion his moral approval or the reverse, according to the strength or weakness of the temptation? Is not the weakest temptation which results in act, as strictly as the very strongest, the sufficient reason of the act, and in so far forth the excuse of it? Two temptations, a strong and a weak respectively, having induced act, does Mr Mill really suppose in the strong temptation any compulsory *power* to induce its act, which did not also reside in the weak one? And why talk of strength or weakness of temptation? These phrases have only meaning in relation to the strength of antagonist impulses, a strength severely predetermined like that of the temptation itself. The question of the result can plainly no more be a *moral* one, than if it simply concerned the tilting of weights on a balance. It is impossible a writer should enmesh himself in a net of more fatal inconsistencies.

Recurring to Mr Mill's—" Yes—if he really ' could " ' not help ' acting as he did; that is, if his *will* could " not have helped it "—it is important to note that from this remarkable deliverance Mr Mill should seem to consider that the man is one thing and his *will* another, with some stirring of a separate life in it, whereby what the man is unable to accomplish, the *will* of the man may. This is in effect somewhat as if we said that a feat which a man had failed to do, giving him his whole body to it, he was likely to achieve with his leg. This notion of a distinction between the *man* and his *will* seems to us a very helpless one; and it is obvious to us to recognise in it one main source of the confusions which have hitherto clouded the discussion of a subject not in itself insusceptible of reduction to the clearest logical issues. The *will*, conceived as other than the *ego* itself in act, as something interpolated *between* the *ego* and its act, is the merest metaphysical phantasm that ever bred maggots of bewilderment in the brain of a philosopher entertaining it. We defy any man who will take the trouble to ascertain and define what he means by the *will*, to conceive and define it in the concrete, otherwise than as the veritable man himself, operative in his act, termed volition. It seems certain, at least, no other definition can be given of it, which would sanction our speaking of the Will as an *agent*; and plainly Mr Mill here so speaks of it. Now *does* Mr Mill in this passage mean to identify the *man* and his *will?* It would be to insult Mr Mill to suppose this; for if so, what in effect do we find him writing? " Yes—if he really ' could not help ' acting as he did; " that is, if he could not have helped it "—and with equal pertinence he might have gone on to write—" that " is, if he could positively not have helped it; that is,

" if he could positively *not* have helped it," and so over some pages, which, however Mr Mill might, in writing, have attributed importance to them, could really have had very little, as merely reiterating idly the one first exhaustive clause—" Yes—if he really 'could not help' " it." Now, as we cannot suppose a man like Mr Mill capable of thus accumulating identical propositions, we must suppose him to conceive of the Will as an active entity, distinct from the activity of the man. One fancies with some amusement the scorn too lofty to care to express itself, with which Mr Mill would find such a view seriously imputed to him. And seriously we do not of course impute it to him. But then the question recurs —What *does* Mr Mill mean? It may be he speaks of the Will, as a mere *mode* of the man's activity; but if so, again he speaks inaccurately; for plainly his language implies that the Will is not a mere *mode*, but an *agent*. " If his *will* could not!"—if the Will be a mere mode of the man's activity, how distinguish this from the first simple " If he could not?" In the mere *could* given at first, we have all that is meant in the *will*, afterwards given in explanation. On this ground, again, Mr Mill is convicted of accumulating clauses without meaning. Thinking—to illustrate easily—is also a mode of man's activity; and what should we say of a writer capable of gravely delivering himself thus—" If he had " only thought a little, before he acted—that is, if his " *thinking* had only thought a little?" We should say he was the sort of writer who had better not write at all, and decline to further concern ourselves with him and his speculations. And how a form of statement which is obvious nonsense, when used of an intellectual mode of activity, should become wisdom as used by Mr Mill in regard of the voluntary mode, some ingenious gentleman may perhaps be able to explain; but only a very thoroughgoing admirer of Mr Mill is likely to make the attempt. What then—the reader naturally may wish to know—*can* be Mr Mill's real meaning? And the answer is not far to seek. By Will here, Mr Mill means—Free-will—in virtue of which only, as we have shown, could a man in *any* case " have helped " his act, as Mr Mill here plainly implies that in most cases he could. Here, as elsewhere, Mr Mill can only attain a seeming success in his argument by filching a use of language in strictness only competent to his opponents. What, on Mr Mill's principles, must be stated as the essential distinction between a voluntary and an involuntary act? Simply, we opine, that in the first we are *conscious* as active; in the other, not so. Any consciousness of *power* in the origination of the acts called volitions, Mr Mill expressly says he does not find in himself; and if he *did* find such a consciousness, his " Theory " of Causation," which with rigour excludes all *nexus* between the cause and its effect, would compel him to treat it as nugatory—as " a so-called consciousness " and " source of delusion "—his convenient phrases, as we have seen, for facts of consciousness even before admitted as such, when, by some change in his point of view, they are brought into conflict with his "theory." But others profess (and so far as we can interrogate *ourselves* we concur with them) to be conscious not only of acts, but of the immanence in these of a power, or originative energy, as in some inexplicable way a concause of them, along with their other motive causes, not by any one denied;* and of *this* it is that already in the frame of language, which is shaped by primal human instinct, and not by the subsequent conceits of philosophers, the term *will* is used as a convenient

---

* It has always been a favourite line of argument with Necessitarians to represent the advocates of Freedom as denying Motives or Causes. But whatever pertinence the argument may have had against previous advocates of Freedom (and at one time it had at least a *quasi* pertinence) to urge it against Sir W. Hamilton, or those who in some more or less modified form adopt his views, is to be either mistaken or disingenuous. Yet this we have seen Mr Mill do; with the facts staring him in the face that Hamilton asserts Free-will "inconceivable," on a ground of fact specially insisted on, that "the will is determined by "motives," and throughout his "Notes" on Reid, is merciless to that specially pet philosopher in his reasonings seeking to establish that the Will may act "*without* a motive." Doubtless if we try to *explain* Freedom, we shall inevitably find ourselves driven on some such solecism as this. But Hamilton does not try to explain Freedom; he tries only to show that though clearly seen to be inexplicable, it is yet, as such, to be *believed*. Wherefore, the denial of motives—absurd in any one— would in *him* have been gratuitous absurdity; so that even if some casual expressions could be cited, which might seem to convict him of this, it would be either unfair or unintelligent to regard them as anything but mere inadvertencies.

synonym. Nobody who does not in this sense use it has any right to it at all in the argument. Certainly Mr Mill has none; *this* we undertake to prove; and, as it chances, the proof of it is easiest precisely at that central point of the discussion, success at which against Mr Mill must be held utterly decisive.

Mr Mill comments with some severity on Mr Mansel for his "mistake in thinking that the doctrine of the " causation of human actions is fatalism at all, or re-" sembles fatalism in any of its moral or intellectual " effects. To call it by that name," he says, "is to " break down a fundamental distinction." It is our hope some few of our readers have by this time begun to surmise that this "fundamental distinction" exists only in the minds of Mr Mill and such of his more faithful disciples as will go with him anywhere on trust. And if this has not already been made sufficiently obvious, it may presently be made somewhat more so by Mr Mill himself, when he proceeds to define the distinctions of contrast between Fatalism and his own doctrine. "Real Fatalism is," he says, "of two kinds—" Pure or Asiatic Fatalism—the Fatalism of the Œdi-" pus,"—with which here we are noway concerned—and another kind thus described:—"The other kind, " Modified Fatalism, I will call it, holds that our actions " are determined by our will, our will by our desires, " and our desires by the joint influence of the motives*

---

* In the next page, "a motive being a desire or aversion," (for this

"presented to us and of our individual character; but
"that our character having been made for us, and not
"by us, we are not responsible for it, nor for the actions
"it leads to, and should in vain attempt to alter them."
(Inasmuch as the notion of contingency clings to our
conception of the future—should in vain have attempted
to alter them—would perhaps be preferable; but we need
not split hairs so impalpable.)

Now *this*, we contend, is as close and accurate a
definition of Mr Mill's own *real* doctrine as could readily
be put in words. Mr Mill himself cannot see it so:
and he goes on to define, in contrast with it, the views
he imagines himself to hold. Surely, if *here* he can be
caught tripping, he must needs go down so heavily, that
his steadiest backers will see the sponge had best be
thrown up for him.

"The true doctrine of the Causation of human ac-
"tions maintains, in opposition to both, that not only
"our conduct, but our character, is in part* *amenable*
"*to the will*, that we can, *by employing the proper*

clear and every way unexceptionable definition, we rather think, the
subject is indebted to Mr Mill, though it is pointed at by previous
writers, as in Hamilton's "Mental Tendency,") by motives Mr Mill
here of course means "objects of desire"—"external motives," as he
has elsewhere inaccurately called them.

* "In part." As Mr Mill has not cared to say in what part, it is
almost needless at all to advert to this. The other influences glanced
at are, of course, those of circumstance, furnishing objects of "desire
"and aversion;" but plainly these may be dismissed. Coleridge's
famous aphorism—which has seemed to some so important—"The

"*means*, improve our character; in other words, we are under an obligation to seek the improvement of our character."

Observations these highly edifying, doubtless, but, as used by a Necessitarian, or unconditional Causationist, conclusive of his mere bewilderment. Mr Mill could not, if he tried it a hundred years, show that by " our conduct is amenable to the will"—unless, indeed, it be Freewill—he means anything more than that our conduct is in fact our conduct,—a remark undoubtedly true, but not philosophically profound. As to " our *character* is amenable to the will," it refuses to become evident to us that, in so stating it, Mr Mill has any meaning whatever; indeed it is evident that, as a reasonable creature, he ought positively not to have any. The *character* is amenable to a *will*, which Mr Mill expressly maintains to be the *mere creature and slave of the character*. It is a clever creature, it seems —as clever as the great Hegel* was—and goes on creating The " wise child" of the proverb seems here a little outdone; but in a child wise enough to positively know its own father, it is perhaps an addi-

---

" man makes the motive, not the motive the man"—a motive urgent to one man being no motive to another—though in strictness it takes us no farther, takes us at least thus far, that it enables us with perfect confidence to eliminate Mr Mill's "in part" here.

* Hegel—one of whose lectures to his students commenced thus: " Gentlemen, by your kind permission, I will now proceed

tional point of wisdom to refrain from any frantic attempt to pay him the return compliment of begetting him. In the will, as a necessitated product of the character, there can plainly be nothing which was not previously in the character; as the character is, so *must* be the *will*: given an evil character, we can only have an evil will; and in this evil will, Mr Mill absolutely maintains a power to *determine itself* to react on the character for *good*. Mr Mill, it seems, is not the Necessitarian he supposes himself, but a wild advocate of Freedom. That "we can, by employing the proper means, im-" prove our character," may be fitly considered in the light of this previous passage about volition :—" Direct " power over my volitions, I am conscious of none. I " can indeed influence my own volitions, but only as " as other people can influence my volitions, by employ-" ing the appropriate means." How other people might influence Mr Mill's volitions, we understand. If Mr Mill were a bad boy, they might scourge him into amendment, or bribe him to it with an apple-tart. If, in this figure of a schoolboy, he should write as we here find him doing, the scourging he would hardly escape. For how, except as a free agent, is Mr Mill to " influ-" ence his own volitions?" He is to do it by " the " employment of appropriate means." But without exercise of prior volitions, he will probably find it not easy to employ these appropriate means; and as over these prior volitions he has " no direct power," previous appropriate means must be employed by him; and so on as far as we please to carry it. How ever is Mr Mill to get commenced with his operations upon *himself?* He must go a good way back. He may go back to his birth, if he pleases; we will even, on the larger latitude suggested by Tristram Shandy, allow him some few months farther. Again, *what* are the " appropriate " means" which Mr Mill proposes to employ ? They can be nothing, of course, but appropriate *motives.* Such motives Mr Mill is therefore to furnish to himself; and "since motives are desires and aversions," it is *these* he will have to provide. But the complex of desires and aversions, active and latent in the character, is at any given moment a severely determined quantity; so that unless he can freely originate the *new* desires and aversions desiderated as "appropriate," we see not how the deuce he is to get them. Mr Mill, who in his modesty was sceptical as to his " direct power over his " volitions," in the matter of his desires and aversions asserts for himself a power of spontaneous generation unlimited. Again, as we before said, he is not the Necessitarian he supposes himself, but a frantic advocate of Freedom.

Now, if Mr Mill, scouting Freedom, on the one hand, yet shrinking from Fatalism, on the other, cannot state his intermediate doctrine without coming to such dreadful grief as we see, the inference is scarce to be evaded that there is *no* such *via media* as he would indicate, or at least that as yet he has failed to find it. And if Mr Mill has, up to this date, been unsuccessful in finding it, we may fairly set aside as a remote one the chance of his success hereafter. Wherefore, the alternative pressed upon him is, whether he will logically go on to Fatalism with all "its moral and intellectual effects," or *non*-logically retreat to Freedom? Non-logically, we say, not *il*-logically; for without any compromise of his logic—whatever might be said of his consistency—this line of retreat lies open to him. In order to make it available, Mr Mill has simply to admit, that as human reason is not necessarily the measure of all things—and Mr Mill in the frankest way would admit this—there may be questions which more or less *transcend* logic, and accept the surmise of his whilome antagonists, that of such questions this concerning Freedom is one. And in perfect consistency with his own principles, Mr Mill might elect to do this. The Will, as we before said, cannot in any concrete* sense be otherwise defined than as the *ego* itself in act, the veritable *persona* of the man. And has

---

* Of course, if we chose to define the Will in some more abstract way, as a mode, function, or the like, and having done so, are careful never to use the word save severely as so defined, to this there can be no objection. But it would probably be found that on these rigorous terms, we should never in this dispute hear mention of the Will at all. And perhaps were there never more to be mention of it, some needless confusions might be avoided in the future stages of the controversy—not unlikely to last with the world—as to whether or no there be in man—not, let us say, a Will of any kind, but—a Free force—an ability to act otherwise than as he does act.

Mr Mill, with his little plummet line of logic, so sounded "the abysmal deeps of personality"* as to be able to announce with assurance that they have yielded him their ultimate secret? Mr Mill himself makes no such pretension. Face to face with the Ego he admits himself, as all men must do, in the presence of an inscrutable mystery. His attitude assumed to this mystery is not one of any such awe, as that of some other thinkers —or dreamers, as he would prefer to call them—who find in it the type and guarantee of a mystery more high and sacred. Rather, he seems to regard the *Ego* more or less in the light of a *bore*, in its declining to come and be included under the forms of his "Logical System." And along with this natural and excusable disgust, there is almost perhaps to be read in him a trace of quite ingenuous *surprise* at this contumacy on the part of the Ego—this stupid unreasonable Egotism, as it were, and utter disregard of the claims of accurate thought. Nevertheless, though his Logic can give no account of it, Mr Mill admits—or as good as admits—the *Ego;* which is really to be recognised in so great a Logician as a trait of honourable candour. Apart from its positive value, as this may finally be determined, Mr Mill's application of what he calls the "Psychological Method" to the

---

* "God, before whom ever lie bare
　　The abysmal deeps of Personality."
　　　　　　　—Tennyson's *Palace of Art.*
Vide Arthur Hallam :—"God with whom alone rest the abysmal secrets of Personality."

phenomena of Matter and Mind, will probably by competent judges be ranked as by far the most important contribution to Mental Science which has for many years solicited attention. Postulating "the human mind "as capable" of Sensation, Expectation, and (we venture to add) Memory, as of this last the necessary correlate*—postulating, that is, the human creature, as Shakespeare defines it, "looking before and after," Mr Mill proceeds to construct for it an ideal world—with what amount of positive success we could not undertake to say. Mr Mill's ideal world seems a very good ideal world; we cannot see any great objection to it, except the frivolous and stupid one advanced against others of its class, that it is not the least like the real world. But, admitting its entire excellence, as the postulated human creature, invariably on its advent, finds a world ready made† to its hand, we fear it may not sufficiently appreciate the pains here taken by Mr Mill to find it an ideal outfit. Having thus constructed an ideal world of mat-

* Of course, in expecting, we must expect something, the elements of which at least are given in our previous experience; to expect nothing is plainly not to expect at all; our most fine-spun vacuous shadow of the future involves some shadowy dream of the past.

† To be candid about it, this is just what Mr Mill denies. His notion is that the human creature weaves its world for itself pretty much as it does its breeches. Perhaps it might also be maintained that previous to weaving its breeches, it weaves its own legs to be covered by them; but meantime Mr Mill does not quite see his way to this. The reader will excuse these levities. "The end and the beginning vex," so that they tempt one at times to indulge in such.

ter, Mr Mill with equal ingenuity proceeds to fashion one of Mind; but after having done so quite to his own satisfaction, he finds he has made a slight omission. In his ideal world of Mind, he has forgotten to include the Ego; and this pestilent Ego, it seems, when he remembers and seeks to include it, is found contumacious and intractable. With its "*mystic* faculties of Hope and "Memory," (to quote Mr Carlyle, who, with little Metaphysical aptitude, has vivid Metaphysical insights,*) involving as they do the "paradox of something which "is only a series of feelings, aware of itself as a series," the Ego is too much for Mr Mill, and has to be included strictly *on its own terms* as wholly to thought "inex-"plicable." It can scarce escape the attention of even a careless reader of Mr Mill, that this Ego—with its awkward gifts of Memory and Expectation—which turns up as an "Inexplicable" at the close of his speculation, is the very postulate from which, as we saw, it proceeds. Whether a speculation which assumes as its necessary postulate an Inexplicable, can truly be held to *explain* anything whatever, we are not here concerned

---

* Mr Carlyle did not always despise Metaphysics as he now does, and his earlier Essays include a good deal of exposition of the doctrines of German philosophy. This at the time had its value; but looked into now, much of it seems jejune enough as the product of a mind so powerful, seriously directed to such studies. On the whole, perhaps Mr Carlyle did wisely at once for himself and his readers, when, deserting this field for ever, he went forth on his grand crusade against the general ostriches of creation.

to inquire. But Mr Mill's admissions as regards the Ego seem pertinent to our present purpose. In presence of the Ego we are "face to face with a final inexplicability"—with an "inexplicable fact" which yet we are compelled, as such, to "accept." Now, if Mr Mill can only be prevailed upon to go along with us in identifying the Will with the Ego, it is plain our dispute with him is narrowing itself. And to this identification he is almost committed in his every way unhappy—"Yes, if he 'could not help' acting as he did; that is, if his *will* could not have helped it," which either means *that*, or nothing. But, discarding the word Will altogether, if the Ego is admitted, as we see it, an utterly mysterious entity, may we not logically surmise in it properties more or less mysterious? Would it not, in fact, be wildly illogical to do otherwise? The "inexplicable" Ego may, for aught we know, be a complex of some fifty or five hundred inexplicable attributes. The claim to be allowed the assumption for it of only *one* little attribute of Freedom, admitted wholly "inexplicable," seems, therefore, an entirely modest one. And it may well seem, also, a reasonable one, if only by assuming this attribute we can rationally continue to believe in the validity of our moral nature. That our logical faculty rejects it can surely—this *if* being given in the affirmative—be no good reason that *we* should. Even were the human intelligence a mightier matter than it is, man is not simply an intelligence; the mere instincts of sense apart, which he shares with the brutes beneath him, the intellect on which he so piques himself, as the instrument of his vaunted science, is really the paltriest of his possessions; inextricably interworked with it he has beautiful emotions and affections; indestructible yearnings attached to these which oversoar the mists of time; passions in which, when the mortal taint in them is deepest, some heavenly longing yet lingers; hopes which from their own ashes recreate themselves, and vaster dreams, prophetic to him of fulfilments of unknown desire; and, central to this whole mystic apparatus of spirit, the keystone as it were, of the arch, which keeps it all from flux and ruin, there is given him in consciousness a system of moral beliefs, and what has well been called "an infinite law of duty." If in order to conceive of this as a reality, and not a mere deceptive dream, he must postulate as real the Freedom which he also finds in consciousness—and that he *must* we have abundantly shown—we confess we see not on what grounds any Logic not wildly arrogant can refuse to concede his postulate. The moral Consciousness not less than the Logical one being admitted as a valid fact of human Nature, by what right can this last deny to it the assumption (supposing it *were* a mere assumption, and not really a *datum* of consciousness) which is needed to constitute its validity? By no right which does not involve in its exercise a denial of that validity, and a claim on the part of the logical consciousness to arbitrarily suppress the Moral one; a claim which the Moral consciousness might with precisely the same right retaliate as against the Logical. But, happily, we need not thus set these different parts of our nature perpetually together by the ears; having convinced ourselves that each authentically *is* an authoritative factor of our complex being, we may accept the authority of *both*, believing in Necessity as all-including on the one hand, in Freedom as mysteriously limiting it on the other. To deny so plain a deduction of the intellect as the first would be to " put out the eyes of our " mind;" but not the less is it allowed us to hold fast to the other as an implication of our moral consciousness. That we cannot logically *reconcile* these beliefs, accepted each on the evidence which seems appropriate to it, is really no proof of the necessary falsity of either, but simply, as before we said, one more illustrative instance of the limitation of our logical faculty. That propositions which, as our intelligence is now constituted, obstinately remain irreconcilable, may yet somehow, beyond the sphere of that intelligence, admit of being reconciled, seems no such extravagant proposition. Moreover, to say that Necessity, as determining human action, cannot be harmonised in belief with Free-will as in some sense and measure* *self*-determining it, is really no

---

* In some sense and measure. If we conceive of Freedom and Necessity as mysteriously co-existing in the voluntary acts of rational creatures, it is plain we can only so conceive of them in degrees and variable proportions. It would be too palpable an outrage of reason

more than to say that Free-will is a mystery, as which it is expressly announced. We expressly announce the Will a mystery, and necessarily so, as one with the inscrutable human personality, by Mr Mill "accepted," as we saw, while admitted wholly "inexplicable." And does Mr Mill really consider that in utterly mysterious and inexplicable entities everything should be plain sailing for him? Should Mr Mill care at this date to take up with Freedom, there seems nothing to prevent his doing so, and keeping his Causationism also. But this he is not likely to think of; wherefore, as his search for a *via media* has sufficiently been proved abortive, the only

---

to say that a man driven by a strong motive is in his act *as* free as he who is the subject of a weaker one. On the other hand, to say that a starving wretch is no more necessitated to eat than a man who sits down to supper after dining heartily an hour or two before, is the same thing stated from the other side. That he *is* no more necessitated, speculatively, we must assert; for, as we have abundantly seen, it is the clearest deduction from the law of Causation, as applied to them, that, in relation to acts induced, the most absolutely tyrannous motive is precisely as the most absolutely trivial one, each being the sufficient determining cause of an effect inevitable as determined. But, practically, it is on all hands recognised that a strong motive necessitates, and thus excuses action, as a weaker one does not. We must thus conceive Necessity and Freedom to coinhere in human action, as variable quantities reciprocally limiting each other; so that, whilst every action is in some strict sense necessitated, it is still in some such sense *free*, as permits us to ascribe to it a moral quality. This is, of course, to logic absurd; but, as a rough-working conception of a mystery *beyond* logic, the trustful acceptance of which is indispensable to a rational belief at once in the results of science, and the validity of our moral nature, perhaps it may pass for the nonce.

alternative left him is a plunge into the abyss of Fatalism, with "all its moral and intellectual effects."

As to the moral effects of Fatalism, they have incidentally been noted, on their speculative side, in the discussion which had for its object the exhibition of Mr Mill's Causationism as only distinguished from that doctrine by certain unmeaning refinements. That a Fatalist can only continue to believe in the reality of moral distinctions, at the expense of his logical consistency, it would be waste of time to try to prove further. In any scheme of thorough-going Causationism, Necessity, or Fate—the distinctions are merely verbal, save only as in the word Fate, a positive extinction of Freedom by a hostile power seems asserted, as in the two others it *needs not*, as we saw, be held to be—Morality, if still considered to exist, can only be recognised as an illusion. The Moral judgments which attribute merit or demerit to conduct, whether in ourselves or others, must be only so much hallucination: and however they may still usefully for a time continue to delude the vulgar, can plainly have no place in the creed of an advanced speculative intelligence. When a young woman, without having qualified herself in the legal manner, has inadvertently added an item to the population, the easy phrase which extenuates her lapse as "a misfortune" is not so properly a charitable concession to feminine frailty, as the dictate of an enlightened philosophy; and when, finding her babe inconvenient, she incontinently plucks its head off, this is simply one little misfortune the more for her. A Palmer, a Rush, or a Pritchard is no more an object of legitimate moral indignation, than Howard or Mrs Fry of an intelligent moral approval. The relations of their actions to utility, of course, remain; on which ground of distinction we may properly encourage the one class of persons, and discourage the other by hanging them as good riddance, and some hint of a warning to a public, supposed to be amenable to motive. But *moral* distinction there is none; and as far as any *desert* in the matter may go, to canonise a Howard when dead is no more a rational proceeding than it would have been to hang him whilst living, or than it would now be to canonise Rush. These results are curious; but they are logically involved in a system in which—Freedom denied —every human action is conceived of as simply the last link in a chain of pre-arranged causational sequences, and so not possibly to have been altered or evaded on any easier terms than a rupture and re-arrangement of the whole vast chain from the beginning. With such a conception, inasmuch as it may include the fear of Punishment, as one motive among others, responsibility *simple* may consist; but with Moral Responsibility in man, which implies good and evil desert in conduct, as the subject of praise and blame, of righteous reward and punishment, it is plainly and for ever incompatible. But we need scarcely so iterate a statement already sufficiently enforced.

It may be said, however, on the analogy by ourselves suggested, that admitting our moral beliefs illusory, this mode of conceiving the matter is a mere curiosity of speculation, like the whim of the idealist in his denial of the existence of matter, and could never any more than that be fruitful of practical inconvenience. Idealists comport themselves precisely as others; no more than a Natural Realist does your Idealist plunge over precipices, or (when sober) try absurd conclusions with lamp-posts. The inference is reasonable, it may be said, that people would analogously respect their moral beliefs in practice though they might speculatively have come to reject them. There is perhaps a certain distinction fairly to be noted between the two cases. The idealist, as we understand the matter, cannot in any accurate sense be said to deny the existence of an objective external world; he denies it as the sceptic—atheist so called—is said (by certain persons in pulpits) to deny the existence of a God; he says that the *proof* of it is deficient, that on a critical analysis of the so-called "natural belief" he does not find it amount to proof. But that there cannot possibly be a real external world to which his "ideas" may be conformed, the utmost hardihood of idealism will perhaps scarce go the length of asserting. On the other hand, the argument from Necessity, in regard of our moral beliefs, amounts to a positive rational negation of them, not merely to a sceptical questioning as to the nature and sufficiency of the proof. Shorn of the attribute of Freedom, and subjected to a blind law of Causation, man cannot possibly be a moral agent, and morality *must* be for him an illusion.

> "Roll'd round in earth's diurnal course
> With rocks and stones and trees,"

and acting under laws as inexorable, and as utterly beyond his control as those by which he is thus sped round in space, no more than to a rock, a stone, or a tree, can we rationally attribute to him moral qualities, the subjects of praise or blame. Dismissing this distinction, which, even if admitted—and perhaps there are idealists sturdy enough to dispute its validity—has plainly no very great relevance, there are other distinctions between the cases, which seem somewhat more to the purpose. The idealist, unless also a madman, is under no temptation to run against walls and lamp-posts in the interest of his pet theory. This, he is already convinced by what Mr Mill would call " a complete induction from ex- " perience," though pretty as a speculation, breaks down in its reduction to practice—the penalties sure to be exacted by the wall in his butting his head upon it (supposing him wild enough to dream of such a thing) being very sharp and immediate. The physical world allows us no liberties taken with it. But with the moral world considerable more licence is permitted us; we are happy if daily we do not with full purpose run up against some part of *it;* and we are under hourly temptation to do so, despite of the modified penalties exacted. It seems plain that to decisively announce, and in some sort believe it an *illusion*, might by so much increase this temptation, already too strong for the best of us, even whilst revering it as a reality. On which ground of distinction it would be utterly absurd to reason on the analogy of the harmless idealism of the Berkeleian that this Moral idealism, so to call it, would not be found in practice to issue in baneful results. We are sure Mr Mill at least would not so reason. In his moral recoil from what he dreads as the deadly result of Fatalism, it is obvious to recognise the ground of the indecision which here we have seen clouding an intelligence, elsewhere so clear, trenchant, and conclusive. Repeatedly in his " Logic " we find him admitting the " *depressing* effect " of the fatalist doctrine," and deploring as " humiliat- " ing to pride and paralysing to desire of excellence," the too frequent, and—as he tries to show with such success as we have seen—utterly mistaken identification of this with his " true Doctrine of Causation." Nay, so sharp is his sense of these tendencies, that the doctrine of Freedom, though holding it baseless as a dream, he admits to " have given to its adherents a practical feeling, much " nearer to the truth than has generally existed in the " minds of necessarians," and to have " fostered a " stronger spirit of self-culture " than the opinions to which it is opposed. The candour of these admissions is admirable, but of course they come easy to Mr Mill, who is really, as we have seen—though without as yet, being aware of it—himself a very ardent advocate of the doctrine whose praises he thus celebrates. Now, if Fatalism is admitted to induce these negative results of depression, &c., even in minds (from this point of view it is that Mr Mill writes) which do not press it to its logical result as abolishing moral distinctions, it seems plain that in minds which clearly see it to issue in this, it might readily enough be developed into dire forms of positive evil. And no man who reasons with the least strictness can fail to evolve for himself this result of the doctrine; having done which, he can only, on the ground of logic, regard our current Moralities as a form of superstition, useful, perhaps—as the Christian religion is admitted still to have its uses by many who for themselves will have none of it—but not otherwise entitled to the respect of an advanced emancipated intelligence. Of the practical issues we had proposed to treat at some length. First, as to how such an intellectually emancipated person would be likely to comport himself in a world of the yet unemancipated; and again, into what sort of world, a world wholly emancipated might in some little time be developed, it might be not without interest to inquire. But the inquiry would be rather more difficult and intricate than we can fancy it might seem to the perfectly "well-constituted mind," and, in any case, is perhaps a too merely *curious* one to be worth taking any great pains with. Moreover, the subject, in its very nature, is dreadfully beset with respectable moral and religious platitudes; and, unawares, one might find one's-self with a deep air of wisdom promulgating such, as the latest important " discoveries" in this particular field of thought. As it seems, on the whole, desirable to avoid this, a swift confused outline must suffice.

Precisely according to the decisiveness with which we recognise moral ideas as *illusions*, it is plain we get rid of them as *motives*. Suppose this decisiveness complete; the " internal sanctions " of conduct are made away with —*conscience* no longer exists, to " kindle or restrain." The " external sanctions " remain, but not quite as they were. That important section of them which rests on the *moral* approval or disapproval of our fellow-men has, of course, evaporated—it has absolutely, so to speak, evaporated in the emancipated world—relatively in the emancipated individual—on the obvious ground of the extinction in him of the special sympathy. Also, in the emancipated world, the other remaining "external sanctions" might come to be much more languidly enforced than as now they are, in virtue of a deadly moral indifference, which—even in the supposed disappearance of *all* virtue—would be nearly sure to proclaim itself the virtue of charity. Briefly, the emancipated world would be simply the world as it now is, reduced to its basest beggarly elements; a world in which fear on the one hand, and appetite of some kind on the other, would be the sole admitted motive forces; the Brute world, in fact, which

Mr Mill, as we saw, in his argument substantiating Justice, tried to pass off upon us as the Moral one. A world this, as it seems to us, with some dearth of noble emotion in it—in which only a brute could care to live, and in which, if a *man* were supposed capable of living, the desperate devouring desire of his soul in every instant of his existence would be to turn Armstrong guns upon it; a world, in a word, in which, if any men chanced to linger, the brutes would be sure to hang them, on such excellent grounds of "Justice" as we have seen Mr Mill enunciate. To concern ourselves with the possible or probable doings of such a world would be to "con-"sider it" very much "too curiously." But would or could such a world be on the hypothesis evolved? Unquestionably, we think it would, time sufficient being given. As to the amount of time needed, one would not choose to be specific. Suppose man originally a brute, as, in fact, it is now become fashionable to do; the "moral idea" gets into his head somehow, (*how*, it would be plainly unscientific to ask,) constitutes Society, and in course of some thousands of years brings it to what we see; it would be stupid to think that, suppose again the "moral idea" withdrawn—*presto*, the human world it had constituted would on the instant lapse back into the Brute one. But instantly the *tendency* to so degrade itself would begin to operate in the world, and—give him time—how much we decline to specify—our faith in man is fixed that he would succeed in reproducing the real original Gorilla, so as even to satisfy the strictest scientific requirements of the Professor Huxley of the period.

Now this *tendency* in the species, as supposed, is all that a cautious person will permit himself to announce in the individual unfortunate enough, in taking up with Mr Mill's doctrine, to be a bolder logician than he. It is a tendency, of course, which might in various ways be counteracted; and in all but very desperate cases is sure to be indefinitely counteracted in at least one way. The hypothesis on which we were reasoning may happily, on the whole, be regarded as practically an impossible one. In Mr Mill's remark (*vide* " Logic ")—" A fatalist be-" lieves, or half believes, (for nobody is a consistent " fatalist,) that," &c., there is a ground of cheer and reassurance. But how comes it that nobody can be a consistent fatalist? Had we asked Mr Mill this question when he wrote as above, his reply would have rung out silver-clear: Because of the " feeling of moral Free-" dom we are conscious of." Were we now to put him the question, no doubt Mr Mill could answer, but not, perhaps, in his *tone*, so quite like a silver bell. The bell would be found in the interim to have somehow or other absorbed into it a pestilent alloy of lead. Clearly, if a Fatalist only " half believes " in his Fatalism, it is because he is in what we have seen to be Mr Mill's case, and " half believes " in Freedom. But if a man is at all to believe in Freedom, he had better go in for it entire, pretty much as it is wiser in a waterman to keep the two oars to his boat, than to pitch one of them overboard, and painfully scull it with the other. To conclude of this tendency to moral deterioration involved in even the most merely speculative denial of Moral Freedom, let us cite the deliberate judgment of a man, whose clear title to be heard in such a matter Mr Mill will perhaps not care to deny. We find it thus written by Fichte :—*

"The influence which this Philosophy, (Kant's,) particularly the *ethical* part of it, has had upon my whole system of thought, the revolution which it has effected in my mind, is not to be conceived. To you especially I owe the declaration that I now believe with my whole heart in *free will,* and see that *under this supposition alone can duty, virtue, and Morality have any existence.* From the opposite proposition of the Necessity of all human actions must flow the *most injurious consequences to society;* and it may, in fact, in part be the source of the corrupt morals of the higher classes we so much hear of. Should any one adopting it remain virtuous, we must look for the cause of his purity *elsewhere than in the innocuousness of the doctrine.* With many it is (the cause of?) their want of logical consequence in their actions." (Which

---

* As quoted by Mr Lewes in his "Biographical History of Philosophy," a work in which, in a lucid, lively, and readable way, all the information is condensed that the general reader need desire.

last result, curiously enough, as held to be logically involved in the opposite opinion, is one of the chief considerations perpetually pressed on its adherents by the advocates of the Doctrine of Necessity.)

And now, enough of Mr Mill on Freedom, on which topic his success against the essential doctrine of Hamilton cannot be held great. As to how far he can truly be held to succeed in his attack on the other main doctrines of Hamilton, our knowledge of these at the source is too cursory, and quite superficial to entitle us to form an opinion. Mr Mill's polemic is as pretty intellectual entertainment as any one could desire; and the vein of deferential irony which pervades it—an irony so subtle and skilfully veiled as to have passed with many of his readers for a chivalrous refinement of courtesy,—reveals a turn for pleasantry in Mr Mill, which before we had not suspected, and which has even at times reminded us of that *soupçon* of lurking humour which charms upon the page of Hume. In his argument, unquestionably he often attains a *seeming* success; but in the chapter we have been considering we have found that a seeming success (and no part of his book has been more lauded than this) may prove on a sharp examination to be very far indeed from a real one. Moreover, we have seen *how* he seems to succeed—by being at one time as inconsistent with himself as he tries to prove Hamilton; at another, by misrepresenting the doctrine of the man he is trying to confute. We seem to have heard tell of a gentleman who, in attempting an assassination, effected a *felo-de-se;* also of an ecclesiastical dignitary of the seventeenth century who, having confidently announced to the world that " the fame of Milton had gone out in a " stink," solely in virtue of that feat now lives in the human memory. It would be odd if it were finally adjudged that in reward of his latest performance, only such an unenviable immortality as this could be prophesied for Mr Mill. But unless he has elsewhere been happier than here we have found him, this result seems really on the cards. Let us in candour, however, admit that, taken on this topic of Freedom, Mr Mill is taken at a disadvantage, and even an unfair disadvantage, if an inference of his failure elsewhere is any way severely pressed. For suppose the case we have tried to make out admitted thoroughly established. What does this prove against Mr Mill? Simply that he could not succeed in an impossible attempt. His doctrine, consistently reasoned out, is a purely Fatalistic one—in its essence an unconditional moral scepticism—and he will for ever in vain attempt to combine with it the morals of Freedom. Thoughts go free, we hope; and we do not use the ugly word *scepticism* in the least *in malam partem*. Had Mr Mill plainly set forth his moral system as such, we should rather have respected his speculative hardihood and severe intellectual integrity, than have felt called upon to mete out to him any word of orthodox reprobation. (When Mr Mill, in a remarkable and much admired passage of his book (p. 103), says decisively, " to hell I will go"—

be it far from us to answer—Go then! though we
can fancy that to not a few pious minds it might
seem that, by some such curt rejoinder, the whole
demands of the case were satisfied.) As it is, we
reverse-wise consider that his resolute though pathetically
hopeless clinging to accepted belief exalts him as the
moral creature which in strictness he has no right to
consider himself, far more than his lack of utter speculative
fearlessness can be held to discredit him as a
thinker. We take leave of Mr Mill on this topic of
Freedom with a great deal of the admiration and respect
he so handsomely throughout his book accords to Sir
W. Hamilton.

Postscript.

Vide page 24.

"(If our so called consciousness (before having decided,
"of being able to decide either way) is not borne out by
"experience, it is a delusion; it has no title to credence
"but as an interpretation of experience: and if it is a
"false interpretation, it must give way.)"

On a reconsideration of his argument, we see reason
to think that, in our remarks on this passage, Mr Mill's
meaning is misapprehended. He is probably incapable
of the absurdity imputed to him, of alleging in the
singular instance of a man's acting in *one* way, a contradiction
by experience of his previous consciousness of
a power to have acted in *either* of *two* ways. What he
really means would seem to be that the supposition of

any power in him to act, except in the one determined way, is conclusively negatived by that "complete induc- "tion from experience," elaborately set forth some pages before, which exhibits in human actions a uniformity of sequence, as complete as that which we find, or not finding, assume as certain, in all other phenomena. "This "argument, from experience," Mr Mill says, "Sir W. "Hamilton passes unnoticed." As, according to his own statement, "consciousness is not prophetic," and Mr Mill's argument did not chance to be before Sir W. Hamilton, it is perhaps to be excused in him that he did not take any notice of it.* And if it *had* been before him,

---

* The attempt at a point here is wretched. Of course Hamilton had this argument of Mr Mill's before him. He had it before him in Hume, to whom, indeed, Mr Mill is indebted for all his reasonings on this subject, which do not chance to involve him in blunder. And in fact it might be interesting to show, as one readily enough might, how reasonings which were competent to Hume, as incompetent to Mr Mill, have become in his use of them blunders. Any one who, not having Hume to his hand, chances to have a Shelley, will find in the note to "Queen Mab" on "Necessity," a very succinct and skilful redaction out of Hume, which at every point touches Mr Mill's argument. He will also find in Shelley what he will not either in Mr Mill or in Hume, a pretty sharp appreciation of the results of the doctrine, as "changing the established notions of morality," and, to say nothing of its "utterly destroying religion," leaving "the word *desert*, in its "present sense, utterly without a meaning." As to Hume's performance, admirable in much, it is perhaps in nothing so particularly so as in its dexterous cool evasion of the real or *human* difficulty. He brings himself full in front of it, sees keenly—as he shows in one casual word which escapes him—that it is of no use to attack it, and "refusing "himself" accordingly, effects a theological diversion.

probably he might have thought it required no notice. When, admitting the will determined by motives, he decisively announced that, on that ground, "we can " never *in thought* escape determination and Neces- " sity," Hamilton would probably have considered he had granted to its fullest extent the crushing logical force of the argument derived from experience, however in its statement it might be varied. And undoubtedly the argument is on its own ground irresistible. But it seems to us Mr Mill, as he here applies it, has moved it from its own ground to another, on which, on his own principles, it is self-convicted of futility. " A complete induction " from experience " is indeed conclusive, as against any " interpretation of experience" which is *not* "a complete " induction," and satisfactorily proves it a " false inter- " pretation which must give way." But is our consciousness, or " so-called consciousness," *previous* to act, that of two things we are able to choose *either*, to be called an " interpretation " of our *subsequent* " experi- " ence" that of the two we shall choose *one*? This would be to make consciousness " prophetic " with a vengeance. Yet it seems to us Mr Mill must mean this; certain it is, at least, he cannot mean anything to be held in the least more sensible. For of what *possible* " experience " can this "consciousness" be said to be an " interpretation?" The sole experience which a man can ever have is that he has chosen *one* of two things. And how he could ever "interpret" this plain, unmistakable, and most *simple* fact of conscious experience, into an assurance that he was able to chose the *other* of the two things, we confess we fail to understand. It would be quite as intelligible that he should interpret the experience of having done the one thing into a consciousness that he had done the other; indeed, it would be much more so. To say that we "interpret" our *experience* of having done *one* thing into a consciousness of ability to have done *another*, seems about as wild inaccuracy as it is possible to put in words. We admit, that in the text we stupidly misunderstood Mr Mill; but it is pleasant to find him here on an equally fine line of mistake. Nothing can be more obvious than that the consciousness of our ability to have chosen *otherwise*, must originate *outside* of the "experience" that we actually chose as we *did*. To call it an "interpretation of that "experience" seems almost a joke on Mr Mill's part. It is a consciousness, so-called consciousness or conviction—name it as you will—arising outside of experience, and *conflicting* it may be, with "a complete induction "of experience," but which this "complete induction" cannot be allowed, even at the behest of Mr Mill, so summarily to make away with as a so-styled "false "interpretation." For, in fact, it is pretty plainly no such absurd "interpretation," but a consciousness, and a consciousness not merely "so called," but one incapable of being accounted for except as primitive and underived.

Mr Mill has remarked, as we saw, on the difficulty attending all inquiry as to primitive consciousness, involved in our inability to trace the mental processes of infants. But almost so soon as children begin to speak intelligently, significant hints may be caught from them; and it is not a little curious that the earliest utterances of these small philosophers are in favour of Mr Mill's doctrine of Necessity. Every one in the least familiar with children must have noticed how readily for acts which were plainly voluntary—and this in obvious passionate *bona fides*—the little trembling excuse of "I could not help it" leaps to the lips of the diminutive malefactor of the nursery. It is not without some light sense of awe, mingled with that of amusement, that one thus hears "the afflicted will of poor hu-"manity" pipe out its first small pitiful appeal. The force and obviousness of the argument, derived from the experienced sequences of impulse and act, receives here the strongest illustration of which it seems capable —for the feeling of the little creature plainly is that its *will* was overborne, and that so it "could not help" * doing as it did. Mr Mill might be inclined to suppose this a case in which out of the mouth of a babe is perfected the wisdom of a great Causationist philosopher; but the true and resistless inference from it is really the

---

* In the infancy of the race, so to put it, we find in language an analogy of this, in the primitive meaning of the words, *passion*, *affection*, &c. For *passive* at page 17, of course *passions* should be read.

other way. For—except for some living instinct of Freedom in the child, borne down for the instant by a fact of tyrannous experience,—how comes it that, as reason and self-knowledge are developed, the child becomes incapable of offering such an excuse; and this, despite the cumulative force in every hour of its life, of the "experience" which first suggested it? How, if the "collective experience of life," as Mr Mill alleges, gives evidence dead the other way, comes the child to *acquire* the notion of Freedom, and to feel that it "*could* have helped" its every action? How—the child apart—would Mr Mill explain the *origination* of this idea of Freedom, and the fact that, having originated, it holds its own so pertinaciously? His pet implement of Psychological analysis—Experience and Inseparable Association—seems likely to fail him here, inasmuch as, by his own showing, the total fact of our "experience" gives evidence in favour of Necessity. The idea must, therefore, be *extra*-experiential; and except as an original *datum* of Consciousness, no account is to be had of it. And if this were found to be so, and Freedom as an original *datum* were thus established, Mr Mill from his present point of view would be bound to accept it as *truth*. That Mr Mill, on seeing such a thing forced upon him by his present point of view, would change it and instantly take up with some other, seems merely a matter of course. Probably we should find him returning to the position we saw him take up in his "Logic," and telling us that though, somehow or other, for a time he had lost his consciousness of Freedom, he had now been so fortunate as to recover it, but that, as for the matter in hand, it was really nothing to the purpose, as " noway incon-" sistent with the truth of the contrary Theory."

And again, the validity of a *datum* of Consciousness being anew forced upon him, nothing seems more certain than this—that anew he would lose his Consciousness of Freedom, and label it a "so-called Consciousness," having " no title to credence, except as an interpretation " of the experience" which in its clearly and for ever crying " No" to us is in some incomprehensible way " interpreted" as clearly and for ever crying " Yes." Mr Mill's one fixed principle in the matter plainly is, that Consciousness is to be held valid or *in*-valid, precisely as it may seem to suit the needs of his invaluable Doctrine of Necessity.

AN OCCASIONAL DISCOURSE ON

SAUERTEIG.

By SMELFUNGUS.

DISCOURSE ON SAUERTEIG.

Have readers perhaps heard of a certain Herr Professor Sauerteig? and, if so, what in the fiend's name is their thought of him? Truly, a surprising Herr! of whom, and the abstruse ways of him, one knows not rightly what to think;—strangest agonistic product of a time, surely all too prolific of strangest Gorgons and Chimeras. More singular Gorgon than this Sauerteig the sun does not probably now see. Gorgon of a hitherto unexampled figure—on some sides of him lovely enough, on other sides not so lovely—a terror and perplexity to himself at times, as we rather fear, and surely much a puzzle to poor bewildered persons sedulously eyeing him through this and the other pair of critic spectacles; and earnest, if they could only manage it, to be delivered of some reasonable word concerning him. We can say of this Sauerteig, with some confidence, whatever *else* is to be said of him, that for one thing he has the indubitablest *eye*—inexpressibly important organ, out of which are the issues of life—and is resolute to *glare* withal, rather more than is perhaps needful, to the terror of the more timorous class of persons. *Eye,* perhaps on the whole, comparable to that of the great Mirabeau himself, and which Sauerteig is sedulous to employ *other*wise than the great Mirabeau did in mere winking overmuch at pretty women. Ye heavens! Mirabeau! unparalleled hero-figure! great, greatest! with *eye* which obstinately *would* so wink; except for one, August the physically strong, Saxon man of some energy, to this hour seeking his fellow in *one* indispensable department of human industry! Most indubitably this Sauerteig has an *eye*—unhappily only *one*—and that all too concentrated-intense; stuck also, as we observe, hopelessly into his *occiput*, intent on the far ages mainly, and his own posterior conformations and sitting parts. Sitting parts really rather of the lovely type! Alas! what, if *too* lovely?—all too ideal-aspiring, heroic?—good to be casually glanced at from time to time; to be constantly and sedulously inspected perhaps not quite so good— effect of such sedulous inspection not unlikely to be mere wild rage and disgust, with all sitting parts constructed on a *less* ideal pattern. Of sitting parts of the Sauerteig ideal-heroic species, too close self-inspection may be perilous. For the nature of Ideals is peculiar. To discourse at large here of Ideals, and their divine meanings and uses, might lead us far—probably into very deep regions indeed, whither the British reader— poor blockhead he for the most part—might evince a disinclination to follow. Sufficient, perhaps, in this place, to suggest, that Ideals are of the nature of gin and other stimulant, and behove to be temperately taken. He who cannot take his Ideal temperately shall be severely admonished, and solicited totally to abstain therefrom. Dissipated deep sunk wretches, got dead drunk on their Ideals—hero or other—staggering on all our pavements, wallowing obscene in our gutters, staggering up again therefrom to do mere foul battery and assault on the lieges, ("O ye enchanted apes! flunkeys! owls! ostriches!" other the like foul battery and libel) —*such* poor deep sunk mortals *we* adjudge to be flat nuisances, who, sinking to soft sleep in the gutters, may chance to awake in the police offices. Ideals, alas! chief and even sole blessing of man here below! capable, by excessive unwise use of them, of becoming a considerable curse to him, curse as of fire-gin, and, in fact, the very devil himself. Into the hapless soul of some hitherto eupeptic, comfortably feeding man, let there but suddenly find its way some " divine idea of a pork chop," all actual attainable pork is at once fallen hideous, accurst to him. Beside his "divine idea" of it, continually beaming, *glaring* in upon him, very splendour out of heaven, no pork attainable in mere earthly markets is like to be found satisfactory. On this and the other excellent, highly sufficient, succulent pork chop, he will, with the upturned nose of him, sniff mere hero-scorn and disgust. With such a man, as we compute, the poor pork butchers are like to have hard times of it; obvious that the activity of such a man must with some rapidity reduce itself to sheer wild cursing of the pork butchers. To the butchers decidedly unpleasing; (ugly customer *this!* vilipending *so* our wholesome succulent-sufficient pork chops) and for the poor mortal himself, who must live on pork, surely much a misfortune. Poor mortal likely, we take it, to find himself in no long time somewhat scant of fat upon the ribs of him; growing lean upon his "divine idea;" mortal not unlikely to starve, we fear. Surely a quite unwise impracticable kind of mortal. Palpably diseased unclean pork, deleterious, mere semblance and putrescence of pork, its Ideal all too fatally rotted out of it, no man or Sauerteig shall by this writer be called upon to devour. *Such* pork and putridity, foul, *un*ideal, the ideal all rotted away of it, all men and Sauerteigs shall be called upon by this writer, and even unutterably shrieked upon, to exterminate, conflagrate, sweep swiftly under hatches, and, by all prompt effective methods, abolish from the face of a God's earth, *not,* as we perceive, in the soul and inmost fact of it, constructed upon putrid principles. To decline carrion, and curse and even violently throttle the foul wretch, vending it for human food, thou, O Sauerteig, doest well. But to curse likewise at sound meat offered thee, meat not *pure* ideal, yet supportably so, succulent-sufficient, capable of being wholesomely digested, assimilated by even the more fastidious class of entrails! this, O Sauerteig, is *not* so well, is *ill* of thee, we perceive, O Sauerteig! The poor Sauerteig, with that *one* eye of his, *eye* all too concentrated-intense, stuck strangely into the back of his head, too sedulously superintending the far ages in the light of his own ideal, superfine hero-formations, finds no pork of the present era in the very least to his mind; no cut of it all, alas! will satisfy the ravening soul of a Sauerteig gone wild with his "divine idea." This and the other pork chop of the present day, excellent succulent-sufficient, considering itself doubtless to be *first* chop, the Sauerteig will condescend to inspect and apply his profound philosophic nose to; will confront with his "divine idea," wildly denounce it as a *sham* chop, mere semblance, flunkey, and futility of a chop, and presently, with much imprecation, hurl it back at the head of the pork-butcher. The unhappy Sauerteig! getting rather scant of fat upon the ribs of him, we fear; his "divine idea" not nutritive. For a Sauerteig earnest-fastidious after this fashion, what remains, but that, subsisting himself on a severe *minimum* of the *sham*, mere semblant pork of an accursed "swindler century," he betake himself to the centuries old-devout, heroic, whilst pork yet veritably *was*, and ascertain what chops may lie for him in that direction. Chops *galore* in that direction, and of primest heroic quality; veracious; actual *substance* of meat in them, not semblance and putrid lie merely. This and the other middle age, or other historic piece of properest hero-pork, Sauerteig will, from time to time, produce, exhibit, and infinitely jubilate and glory over, making uncivilest comparisons. *This* chop, alas! however, an all too hungry Sauerteig cannot unhappily *eat*, the "hollow Eternities" having been beforehand with Sauerteig here, and satisfactorily "devoured" it some centuries since. The all too voracious Eternities! rapacious! by reason of whose too prompt forestalling of Sauerteig, our hero-chop can only now be sniffed from afar, divine aromas of it, like airs from Araby the blest, coming to us, wafted through the dim times and spaces, with what slight solace may lie in them for the hungry Sauerteig soul. Superficially it might be judged, that, in this matter, Sauerteig may have ground to complain of these sharp-set procedures of the Eternities, so exceedingly rapacious beforehand, devouring his chops away from him in this rather severe manner. Intrinsically, however, one perceives that solely by its *being* so devoured all away from him a century or two ago, does his chop become radiant, divine-aromatic for him. A chop, the severe *actual* of which cannot now be got at, the Eternities having ravened it up some time since, will be highly convenient for a Sauerteig gone wild with his "divine idea." *Such* chop it is so much more easy to *cook* than the actual foul impracticable chop of a present swindler century. To *cook*, O Sauerteig-Soyer, with thine own "patent inimitable *sauce piquante;*"

late M. Soyer himself, really a poor artist compared with thee in this of the culinary historic. *Such* chop is got partially into the fair region of "the possible;" the Possible, to which the Sauerteig hero-Ideal will briskly proceed to "wed itself;" hero-Ideal most brisk-effective, active, which, once well wedded to the Possible, will speedily contrive with the Possible strange new births of heroism to bless the world—Hero Oliver, considered to be hitherto our supreme feat in the cookery line—Hero Abbot, Samson the name of him, also a culinary performance of some merit—Hero Mirabeau with his *eye*, winking overmuch at pretty women, a questionable figure, but *genuine;* exceedingly *genuine*, O Sauerteig, probably as genuine a blackguard as the planet has seen for some centuries—other miscellany of hero figures, foul scoundrels mostly, cleverly *done* into heroism by applications of our "patent inimitable *sauce piquante.*" Truly, as we said, the Sauerteig hero-Ideal, brisk-effective, active, getting alongside of the Possible, will speedily bless the world with strange new births of heroism. Or, speaking under our former figure (pretty nigh as well ridden to death now as if Sauerteig himself had been astride of it; a Sauerteig, who, once well mount him on a metaphor, may be backed to cross a country with it) was there ever an artist like Sauerteig for the *cooking* of historical pork chops? Late M. Soyer himself, we think, distinctly an inferior artist. Thou singular Sauerteig-Soyer! unsurpassed among men, unsurpassable in this of the culinary historic. Artist really in the high sense; these mere fond imaginations of meat of his, all so wonderfully concreted, visualised, in the conceptive-creative head of him: very actually seeming to *live* for us in the singular cookery books and histories, *actually*, and almost as if they could be *eaten*.

A singular Sauerteig-Soyer, taken in the actual fact, girt with his cook aprons and unutterable culinary wrappages, brandishing his hero-gridirons, and infinitely manipulating with his *sauce piquante* and imaginary middle age pork chops, may perhaps be a figure like few, worth glancing at a little in an occasional way. Latest culinary preparation of Sauerteig, long expected, hungered for, here before us at last, in two stout sufficient volumes, published at the rate of one pound sterling per volume (somewhat severe, O Sauerteig)! may perhaps be worth glancing at in an occasional way. Culinary preparation purporting to be of a certain Grimwold, high-shining heroic baronial figure, of the old King John and Richard eras; " much deserving to be " known; hitherto *not* much known; alas! much *mis*- " known as yet, the very little that we know of him." Poor glimpses of him here and there revealed for us in Monk Chronicle of one Jocelinus de Brakelonda; revealer also of a certain Abbot Samson, of whom readers have heard. *Which* Grimwold, a singular Sauerteig-Soyer, will unutterably proceed to *cook* for us, at the

rate of one pound per volume (severe! O Soyer and Sauerteig)! With slight prelude, and jargoning of the understood sort:—hero-hood! earnest soul! noble life! other the like ineffable cants and jargonings, most peremptorily *not* to be here inflicted on poor innocent readers, Sauerteig, in a really rather clever, by no means quite inartistic way, will treat us as a *whet*, in the first instance, to some life-image and visual presentment of his hero-Grimwold. Presentment passably well done in the approved Sauerteig manner. "Stalwart, high
" hero figure; *steel* figure on occasion; mostly in some
" dubious, uncertain wrappages of buff or the like jer-
" kins, and other middle age ware; somewhat grim-
" trenchant in the looks of him; nose massive, (*valde*
" *grossum et eminentem*, Monk dialect of Jocelinus,) of
" type, as I perceive, high Norman; eyes gleaming out,
" clear-menacing, from under the black bush brows,
" highly capable of *glaring*, if need be, and like enough
" to find need now and then;—a clear decisiveness of
" soul, veracity, earnest valour, looking out from the
" whole man, and breathing from every lineament of
" him;—a highly sufficient man and ruler of men, as
" the outcome of him will shortly convince us." With much to the like purpose, such as some of us may have seen before. A bit of historic portraiture not without merit in its way; slight, not inartistic preliminary *cookery* of Grimwold, and *whetting* of the reader's appetite for him. Judge of our blank bewilderment of

mind, when, turning the page briskly to a new chapter, anxious to make further acquaintance with this interesting hero-figure, we find ourselves discussing with Sauerteig—*what* in the fiend's name does a gentle reader suppose? Adam and fig leaves, we may venture to surmise in a modest way, is *not* what most readers would suppose. By the eternities! O reader, no other; Adam and fig leaves, fall of man; thence downwards by a very slow coach indeed, through Noah, (certain domesticities, incidents here, treated with a free humour, amusing enough, but questionable in these demure times,) Noah! infinite other dreary patriarchs; Hebrew eras; old Roman, old Greek eras; still on, on, till finally we find ourselves, wandering lost creatures, (our high Grimwold, as should seem, gone from us, too probably for ever,) wandering, wandering in thick inextricable jungles of Wends, Kurfursts, Margraves, and the like dolefullest "ghosts of defunct bodies;" still passionately seeking for a Grimwold, and, alas! finding none; no thrice-accursed Wend or Kurfurst of them all able to afford us the least hint of our Grimwold. Ye heavens! it is quite too bad; our hero-Grimwold, in whom we really had an interest, and disbursed two pounds to get news of him a little, rapt away from us *so;* and served up to us here, instead of him, mere disinterred carrion of Wends, Kurfursts, Margraves,—doleful creatures, of interest now to no soul, extinct, unavailable;—available to *thee*, O Sauerteig! for making of thing called book, at somewhat a severe figure;—otherwise for ever *un*available, uninteresting; sole poor interest we could have with them, to get them swiftly shovelled underground again if we could, not without deep execration. Disinterred carrion, O Sauerteig! of mere Kurfursts and the like; plain carrion, actively insulting the nostril, to which *no* cookery could reconcile us. Palpable carrion, O Sauerteig! at the somewhat severe rate of one pound *per* volume down for it! phenomenon which, even in a " swindler century," may be calculated to excite remark. Of a Sauerteig, who, advertising his hero Grimwold to us, finds it needful, after one glimpse given of him, to retire upon " Adam and " fig leaves;" and thence, with extremest tedium, through nameless imbroglios of universal Human History and stupidity, to work downward toward his Grimwold, thus much may be said at least, that he has hit upon a novelty in historical method. Be the praise of originality in the matter, likewise of some audacity, nowise denied to Sauerteig! " Igdrasil, the Life-tree!" shriekest thou, O Sauerteig? as partly we seem to hear thee shriek; " Igdrasil! and how it all *grows*, and, " through all times and branchings of it, is ever mysteri " ously *one!* how the present in every fibre of it does, " in most real irrefragable way, rest upon and relate " itself to all fibres of the past; some understanding of " the past, out of which it flowers and rises, necessary in " order to any wise understanding of the present, &c.

"&c." Reflections, O Sauerteig, scientifically satisfactory to us from of old, yet somewhat, it should seem, of the barren species; on their own essentially rather poor basis satisfactory; distinctly *not* satisfactory to us, bosh to us, balderdash, as regards this present matter; the just rage of us, desperately seeking our Grimwold, (having paid our poor two pounds for him,) seeking, seeking through wastes of mere Wends, Kurfursts—tearing our way through the thorny jungles—lacerating our poor souls and limbs there, not to be appeased, O Sauerteig! by twaddling these poor cants and Igdrasils at us. On the whole, to dismiss this sad Kurfurst business, one feels much inclined, on the head of it, supposing such feat achievable, to *kick* Sauerteig as, to some extent, a *sham* and imposition, and desire him to refund some proportion of the moneys too plainly filched from us.

Praise be to the upper powers, however, if nowise to a robber Sauerteig, making us "stand and deliver" in this rather unprincipled manner; by valour, and human patience, exercise of hero-endurance and faculty to dare and do, one *does* at last contrive—with much difficulty and not without tattered breeches, and thorns sticking in the temper of him—to tear himself, lacerate himself clear of the Kurfurst jungles, and *find* his hero Grimwold again. Pray Heaven only, that, once well found again, he prove *worth* the finding; a hero of moderate respectability, whom, without utter loss of character, we could venture to march through Coventry

DISCOURSE ON SAUERTEIG.

with. Having of old experience of Sauerteig and his unutterable hero-procedures and cookeries, we are not without grave doubts—will be shy meantime of striking up intimacy with this Grimwold, on the mere introduction of a Sauerteig, rather given to consort with scoundrelly persons. A Grimwold who looks rather dubious to us; certificate of character from *other* than Sauerteig highly essential before admitting him to undue intimacy. Sauerteig indeed, nothing doubting, girt with his cook aprons, infinitely manipulating with his hero-gridirons, and due " inimitable *sauce piquante*," cooks busily, with vigour even unusual in him. " Right stuff " of properest hero-porkhood here" iterates the singular Sauerteig-Soyer, cooking; with ever the other dexterous touch of the " inimitable *piquante;*" doubtless will—give him time—*dish up* his questionable Grimwold for us in form truly surprising; prove his Grimwold to be very God in fact, whom let all the peoples worship, or verily it shall be worse for them. Easy for us meanwhile, using our eye in the matter—eye other than the Sauerteig *eye*, held therefore by Sauerteig to be *no*-eye, but ghastly eye-*socket* merely, with *spectacles*,—to see through all lacker of the " inimitable" soused over him. that Grimwold is not the thing at all; is by no means much of a God; is rather the reverse of *that;* and, in fact, to be emphatic about it, as ugly an authentic product of the pit as ever was spued up out of it. For one thing, foully given up to drink; evermore going about, with some quarter cask or so, of *mead*, or other fire-fluid of these epochs, fermenting mere madness in the foul belly of him. For six months at a time goes to bed * in his jack-boots—will rush about at midnight "like a per-" turbed ghost;" and, torch in hand, essay to roast in her bed a high Bertha, his spouse—luckily too drunk to manage it. Shrieks, at times "wildly staring," that " something is haunting him," as indeed is plainly the case. Blue devils are haunting him, blue and very aggravated; gross brute, in fact, seldom to be met with except in mad paroxysm of fiercest *delirium tremens.* ("*Royaller* soul," says Sauerteig once, " I scarce any-" where find record of." Not in my whole extensive miscellany of hero-scoundrels? in a sense we can well believe it.) In which high hero-mood, a model Grim-wold had the misfortune one fine spring morning † to—murder his grandmother, Katie (Katte?) the poor old name of her—*hanging,* with his own hands, that venerable ancient gentlewoman; details of the hero-feat obscure, as culpably admitted by Jocelinus; hero-feat itself happily quite indubitable.

Murder of grandmother, O Sauerteig! not a doubt of it; plainly set down there in Jocelinus, unhappily without detail. Singular hero-feat, which Sauerteig, person in all matters of fact of even exemplary rigour and veracity, will nowise try to suppress—will state quite frankly, gently *cooking* the while; consenting a little to *deplore*

---

* Vol. ii. page 281, for this and the other detail.
† Vol. ii. page 290.

even, in order that he *may* cook, may softly insinuate cookeries. On the whole, Sauerteig will skim lightly over such awkward bit of hero business, treating it in an easy way, not without comic touches. To judge by the Sauerteig cookery of it, it might seem that the murder of one's grandmother was a commonplace sort of occurrence; eccentricity of "the grim man," regretable, not quite defensible perhaps, and yet allowances to be made for it; which blockheads, with no *eye* for the heroic, will be so good as to refrain from over much shrieking at. "Not unlamentable," says Sauerteig, dismissing the subject, "but was not the hero-soul clouded? the great "fact of existence grown for the time *too* great to it, "whence, as we saw in our hero Olivers, hero John- "sons, poor poet Cowpers, and the like, black hypochon- "drias, and wretched diseased *insanity?*" Drunk! O Sauerteig-Soyer! cooking here somewhat too highly; the "inimitable" laid on this time really a little *too* thick. Drunk! O Sauerteig! for some six weeks at a time, the all too *royal* soul that he is, going to bed in his jack-boots; whence, as we have seen in many another loose fish, "hauntings" of him by devils of the blue species, sheer mad rage of *delirium tremens,* and our poor old grandmother to go for it.

Of hero Grimwold in liquor, readers are now in a position to judge. Sober, when by rarest accident you can catch him so, we perceive him to be intrinsically much the same ruffian; the excitement of him indeed less; will now, instead of transcendant exploit upon grandmamma, content himself with discharging across the table, at Grimwold *junior*—likely lad of parts, age twelve or thereby—a soup tureen, of copious middle-age dimensions, slightly fracturing the skull of likely lad; Medicus luckily at hand to cooper it somehow* together again. Hero performance greatly admired by Sauerteig, who will proceed to do pœans in praise of it; Sauerteig much enamoured of the "clear decisiveness, clear steady "insight, manfulness, and, on the whole, veracity," evinced by such a procedure, and will ever and again *congratulate* the young Grimwold, "blest as surely too "few are in *so* serving his apprenticeship to a noble "Hero-Father." Grimwold junior, used to it like the eels, his skull fractured every second day or so, will display, as we perceive, if not gratitude, yet stoicism in the business, and receive his soup tureen with composure which might otherwise surprise us. On the whole, as the reader sees, a hero, too surely of the gross ruffianly type, this Grimwold, and man after Sauerteig's own heart, for whom some skill in the "inimitable" may be needed. On the intellectual side of him a dull block; mass of mere stupidity and dull brute unreason, not even, as sheer unreason, able to give decent account of itself; in the Sauerteig cookery dialect, "man of genius, "strangely inarticulate, *dumb;* the deep veracious insight

---

* Vols. 1 and 2—Nearly anywhere you choose to open, when clear of the Kurfurst jungles.

" of him struggling in vain to *articulate itself*, except by
" soup tureens and the like; poet *without speech*, who
" will polish his stanza by such practical methods as lie
" ready to him; the soup tureen always ready."

For readers interested in this hero Grimwold, and wishing to know more of him and his highly peculiar " mode of existence," we extract from Sauerteig, passage of some length. Grimwold, in great force in it, as will be seen, developing himself in several ways; as family man, and, likewise, in wider capacity of Hero Governor, " guiding the dim populations, and, by all wise valiant " methods, teaching, inciting, and even, if need be, co-" ercing and compelling them to soar heavenwards; in " whom, and his heroic methods and procedures, didactic " meanings may lie for us." The chapter is of much interest, and labelled by Sauerteig,

Hen-Roost—'Ware Poultry!

" Dead waste of night, and under all night-caps in the
" Grimwold household, foolishest dreams in progress;
" suddenly there rises from the Grimwold hen-roost,
" poultry yard, dire pother of the feathered tribes; un-
" utterable multitudinous screeching of alarmed fowls,
" startling the starry silences to some extent, and under
" more than one night-cap, cognisant of it, giving rise
" to speculation enough. Foul *vulpes*, as we guess, at
" work there; with such result as the shuddering dawn

" will reveal. Huge ravage of the Grimwold hen-roost,
" and Cochin-china decimations! Woe of woes! un-
" speakable! sacred immense bubbly-jock, succulent fowl
" of the turkey species, fattening carefully this while back
" for our high carnivals, festivities, *it* too rapt away from
" us, and will solace the coarse entrails of foul human
" *vulpes* unworthy of it! Whereat let the reader of the
" more imaginative turn figure forth to himself as he
" can, the rage of a hero Grimwold, and perhaps a little
" come short of it. A Grimwold nowise indifferent to
" his victuals; with a good hero-twist of his own; a
" sound 'healthy animalism' *(Sinnlichkeit)* the basis of
" him, as of most other men I have known worth much
" in this God's world; to whom sacred bubbly-jock is
" most sacred, the hero-rage at loss of him proportionate.
" Imprecation heaven-high on the part of our hero
" Grimwold! Miserable human *vulpes* (man of business,
" as we should now phrase it) who hast done this foul
" thing, *per os Dei*, shalt thou not die hideously tortured
" for it? The passion of the heroic man is terrible to
" behold, apoplectic. Beautiful beloved Bertha, indis-
" creetly seeking to assuage him a little, is handsomely
" served out for it; is knocked down out of hand;

---

* Goethe—Poet so called of the Germans; supremely great figure to me in old literary *dilletante* days and infanthood; now in mature years getting to look somewhat of a *small* figure; his Fausts and the like, once thought to be great and the greatest, now seen to be *fiddle* merely; our high Hero Goethe himself mere pitifullest supreme fiddler.

"knocked down,—as surely she deserves no less, inter-
"fering in that feminine-indiscreet manner—and after,
"by a Hero-Grimwold with iron boots on, severely
"kicked in the epigastric regions,—beloved Bertha, at
"the time, in a slightly interesting condition. Is con-
"clusively knocked down, kicked in the epigastric re-
"gions—boots very iron-efficacious; snivelling a little
"in the unutterable offensive feminine manner, is told,
"in voice clangorous-stentorian, 'reverberating from the
"'domes,' to 'hold her noise, or a worse thing shall
"'befall her;' holds it; picks herself up as she may,
"copiously bleeding, I observe, merely however from the
"nose; with last little sob convulsive-stifled, curtsies
"submissive, in stately antique graceful fashion; and
"sweeps off to her interior privacies, there to do medita-
"tions appropriate, and what little poulticings may be
"necessary. A man with the true hero-stuff in him
"this, as I perceive! not to be trifled with, idly inter-
"fered with; a right stroke in him when needed, to cut
"short all that sort of thing; the swift decisive valour of
"whom, on this and the other occasion, may amaze us.
"may in many ways have silent didactic meanings for
"us. Few things in a hero Grimwold have been more
"notable to me than this due suppression of his woman-
"kind, a feat so unspeakably difficult. Man of Genius,
"as I always say, strangely inarticulate; *dumb* Poet; a
"high family Ideal in the heart of him, *which*, in such
"rude imperfect methods as lie ready to him, he must

" evermore struggle to express; Poet in a very real and
" genuine sense, who will polish his domestic stanza, as
" we see, perhaps in a somewhat effective manner.
" Truly, a most efficient Captain and ruler of men! Of
" men—*and* of women, O beautiful, beloved Bertha!
" copiously bleeding, as we saw, merely however from
" the nose. Of *women;* a feat so unspeakably difficult,
" even Heroes at times not adequate to it. The sly sluts
" that they are! quasi-submissive, all too insidious-subtle;
" old serpent himself in his best days not perhaps to any
" very great extent subtler, insidiouser; winding us round
" the fingers of them, as if we were worsted from the
" wool shop—'Not to Piccolomini then, dearest?' 'No'
" —in male thunders—No! and even thrice and eternally
" *No!* 'Very well, my own! quite so! of course, Dar-
" 'ling!! *you* ought to know best;' making, O heavens!
" distinctest osculatory effort at him. Osculations in
" progress here, audible, exceedingly nauseous to think
" of; ardencies, amatory movements osculatory and
" *other;* conjugalities, infinite unutterable coo-cooings,
" not here to be minutely specified; and thereafter, as
" one could well foresee, ringing of bell, and John
" Thomas despatched to get opera tickets for us. Alas!
" the quasi-submissive, Dalilah-ish, all too insidious-
" subtle! very Heroes at times not adequate to them!
" Are we *men* then, O wretched mooncalf, being oscu-
" lated upon there, in a way very nauseous indeed to me!
" with authentic hair upon the cheeks of us, with some

" force of God-given Freedom in the souls of us? or
" mooncalves merely, with *rings* set in the silly noses of
" us, to be led hither and thither withal? For *thee*, O
" osculatory mooncalf! I perceive that in this Grimwold,
" there may lie much silent monition. A Grimwold,
" once for all, whom *no* insidiousest Bertha-Dalilah will
" be adequate to lead by the nose; to wind round her
" finger like worsted, and, as if he were wool from the
" wool-stapler, unutterably card and spin. Consider
" him a little, O mooncalf!"

(Not to too impertinently interrupt Sauerteig here, might it not be asked whether, on the Sauerteig notions of heroism, it is necessary to remit the mooncalf so far back as the middle age Grimwold for his lesson? Even in the present deep-sunk accursed swindler century, are there not still some lingerings of heroism " adequate," as he puts it, to these high feats, which he all so lovingly celebrates? " Great men have been among us," and, praise be to the upper powers! still *are*, and shall not yet a while, O Sauerteig! utterly cease from out the land. Did not we, in *Times* newspaper of day now passing, take note of one Tim Mooney, hero-soul of Irish origin, dumb poet, doubtless, in his way too, who, " polishing " his domestic stanza " (with poker) by methods as seemed to us nowise greatly inferior to those of Grimwold himself, methods perhaps even to be recognised as *superior*, could the higher artistic details—unhappily, even by *Times* newspaper, not a very squeamish organ, considered " unfit for publication "—have been well looked into, seen into, was, by horse-hair persons, wholly without eye for the heroic, sentenced to " two " years of penal servitude," as payment in full of his heroism? The unhappy Hero-Mooney! fallen, like Sauerteig, on " an age too late," age of mere valethood, unable to appreciate heroes! Would Sauerteig diligently consider this Mooney, and others of the like, actually now extant among us, though hitherto overlooked by Sauerteig? for really there is much matter in them. A Sauerteig, we fear, deficient in the due breadth of view; a Sauerteig, most erudite-informed, deep in German, and much other fool's lingo, yet plainly unread in the Police reports. Would Sauerteig but address himself a little to *these*, and, considering the nobleness which still lingers with us, strive, in some reasonable loving manner, to adjust himself to the world which he hitherto merely flouts at. Doubt it not, O Sauerteig! pluck but thine eye out of thine *occiput*, plant it in the shining forehead of thee, and *look!* the hero-hood thou so worshippest in thy Grimwold, lo you! it is even *here*, here before us in the Police reports. Would Sauerteig but see fit to betake himself with vigour to the new line of study here suggested, whereof, by due aid of the " patent inimitable " much might really be made—what, if even in two or four stout sufficient volumes, at the easy rate of one pound *per vol.*, some " life and times of Tim Mooney," or the like, true prose epic of the *present* era, for which we have been waiting this while back? And now enough of our interruptions, impertinences, and back, with Sauerteig to Grimwold and his high " ware poultry " businesses.)

" Bertha-interferences disposed of, summarily smit-
" ten aside, as we saw, and a beloved Bertha herself
" swept off to her interior privacies to do poulticings
" and meditations at pleasure, remains that a Hero-
" Grimwold with all speed *do* judgment on foul robber
" *vulpes*, and, in some practical impressive manner,
" preach abroad to the dim populations, the divine
" messages and ' ware poultrys,' struggling in the fire-
" heart of him. And here one bethinks him of the
" judicious Mrs Glass, and her ' first catch your hare.'
" To *catch* thy *vulpes*, O Grimwold; *that*, I perceive,
" will be the first nowise most easy part of the business.
" A *vulpes*, as it proves, most sly-vulpine, and as good
" as declining to be caught. Not a trace to be had of
" *vulpes*. Whole scoundrel populations for miles about
" swept together by swift methods and severest scrutiny
" going forward; with next to no result whatever, *vulpes*
" quite steadily declining to be caught. *Vulpes* for
" certain *here*, but the problem of *catching* him a stiff
" one; as at last appears to a Grimwold, awfully impre-
" cating the while *per os Dei* and the like, a quite blank
" and hopeless one. ' Let justice *be* done,' the deepest
" divine instinct of the hero soul; and, lo now! justice
" is slipping through our fingers and threatening *not* to

" get done. Intolerable to a hero Grimwold, in the
" deep heart of him silently revolving methods. Sudden
" it strikes him, very beam upon him out of heaven
" itself, irradiating the grim visage, shooting out from
" the fire-eyes of him. A gleam of sure insight, sure
" steady glance into the fact, and practicality of the
" matter, which probably may prove surprising to well-
" regulated constitutional minds of these periods. *Vulpes*,
" on the one hand, steadily declining to be caught;
" God's justice, on the other, sternly demanding to *be*
" done upon him; what reconcilement is there—can
" there be? For thee, O well-regulated, red-tapish in-
" dividual! for *thee*, in such case, there *is* none, neither
" can there be. But a middle age hero Grimwold is of
" other stuff than *thou*. Of this universal miscellany of
" scoundrels, (some hundred or two,) certain *this* at
" least, that foul *vulpes* whom we seek is *one*. Swift,
" then, from this miscellany of scoundrels, riddle me out
" some score or so, and, look you, knaves! be quick
" about it, or——. By swift method of lot, as I per-
" ceive, straightway the thing is done; satisfactorily
" riddled out from the general ragged mass, stand twenty
" ragged losels apart there, not looking much as if they
" liked it. These, then, decides our Grimwold, with
" triumph in the grim eyes of him; be these, then, our
" foul *vulpes*. *These*, at stroke of dawn to-morrow,
" solemnly, the Gods looking at us—we will, on one
" sufficient oak-bough, satisfactorily throttle and hang,

"and so conclude the business. Let justice *be* done;
"our divine message of ' 'Ware poultry' preached
"abroad to these dim populations, perhaps with some
"little emphasis. Really a person this with the sound
"stuff of the matter in him, I perceive; man who
"actually *sees*; and, seeing what to do, will promptly
"do it, and no mistake; an original kind of man, and
"withal quick-witted-inventive; his device of hanging
"twenty scoundrels on chance of getting at his *one*
"*vulpes*, not perhaps likely to occur to every one; man,
"above all, who plainly has 'swallowed all formulas,'
"in a way infinitely cheering and satisfactory to me.
"And, O brothers! the gods' message for ever present
"in the old-devout heart of him; present, as in these
"times, new-*un*devout, we cannot even conceive of it—
"' Let justice *be* done '—this, as I do perceive, is a
"thing, in these sad days, much worth meditating.
"Let justice *be* done—surely authentic-divine, and
"summary of all divine messages whatever to us! *be*
"done; if not absolutely, accurately, and finally, then
"approximately, by such methods as may lie to hand,
"of lot or the like; *all* human methods, be it observed,
"*being* in the nature of them approximate. Constitu-
"tional methods now much vaunted, in vogue, accredited
"considerably *more* approximate than these rude-vera-
"cious promptitudes of Grimwold, I suspect; no poor
"mortal ever now throttled at the Old Bailey, but some
"prosperous supreme flunkey sitting on the high places,

"much worshipped by other flunkies, intrinsically *de-serves* it better; flunkey whom I, with these itching hands, if only such blessedness could be granted me on earth, with grimmest gusto *would* throttle. 'Progress,' hitherto, as I compute, for all our hallelujahs heaven-high about it, not very great in *this* direction. Endless considerations pressing on us here shall, for the present, be postponed; let justice *be* done, and high solemnity in progress engrossing our entire attention. Solemnity, perhaps, worth looking at a little—interesting—not unedifying.

"Gay dawn, dewy-bright, and up with it, alert to do the pretty bit of work cut out for him our high hero Grimwold, stern joy in the visage of him grim-implacable. Up with it, also, probably with somewhat *less* joy in them, twenty poor doomed scoundrels, whose singular last-Night thoughts, could we but know them, as we cannot, might be preferable to those of Dr Young. Ranged there, under their sufficient oak-bough, sufficient hemp about the poor necks of them, and overhead on our sufficient oak-bough,— pulleys, or the like, most sufficient. Murder-tackle surely rude enough, shocking to the scientific soul of Calcraft, with its superfine 'patent drop,' but perhaps may *do* the business. Gay dawn, dewy-bright; for lo you! how now, in the far Orient, the great sun, punctual, like a strong man to run his race, comes up in struggling fire! quelling the clouds and night-

"shadows, shooting level on the green earth his victor-
"shafts and floods of yellow radiance; the leaves moist-
"shimmering, and the dews upon the tender herb struck
"all into soft fevers with it. Thrilling the wide air,
"infinite small twitter and piping of glad birds, sedu-
"lous, with gratefullest twitterings, pipings, to welcome
"in the new day; ragged losels, to extent of twenty,
"about to die, looking out into all this, listening to it
"all, too surely for the last time, with feelings of some
"little peculiarity perhaps. Feelings, thoughts, doubt-
"less, in their way, peculiar enough, which, on the
"whole, one would not object to see into a little.
"Frightful is it, O losels? certes, *most* frightful! And
"yet, what if intrinsically even more *strange* than fright-
"ful; the frightfulness, in nature's mercy to us, strangely
"absorbed into the *strangeness* of it. Alas, poor losels!
"dim blockheads, who cannot, even here on the grim
"edge of it, get rightly to *believe* they are to die! O
"heavens! is it not all some black foul dream and un-
"reality, from which we may momently wake up again?
"Losels, much puzzled, as I perceive, not in the least able
"to make it all out; on which, poor ground of stupidity,
"and dim brute bewilderment of mind, they can credit-
"ably enough get through with it, dying to a man *game*,
"as we phrase it in our current speech. As, in fact,
"nearly all mortals *can*, when it comes to that; existing
"they know not whither, for most part, with duest
"stupidity and stolidity. Losels now just on the verge

L

" of it, and last noose in final course of adjustment, sud-
" denly, from the shuddering mass of onlookers, rush
" shrieking maternities to extent of five; shrieking, in-
" finitely ululating, after the unutterable manner of
" maternity in such sad case; cleaving with wild cries
" of them the sacred morning silences. Our children!
" O heavens! our poor children! seems but yesterday
" they were babes upon the proud, glad breasts of us,
" soft-sucking all sorrow away from them; and *now*—
" O heavens! We will pardon to poor maternities, in
" such sad case, some little amounts of ululation. One
" poor maternity in particular, ululating high above all
" others, dashes frantically down at feet of our hero
" Grimwold, and will clutch for dear pity at the knees
" of him, still wildly ululating. Foster-mother of the
" Rhadamanthine man, and on *that* head will shriek
" sufficiently. Her Wat! her poor Wat! her *little*
" Wat, who was dandled on her knees with Grimwold,
" and sucked milk from the same breasts with him! and
" now the dear life to be strangled out of him *so;* and
" *can* a high Grimwold do it? *will* her own Grimwold,
" that once was, have the dead stone heart? To such
" effect ululates and pothers piteous a poor frantic foster-
" mother, clutching passionate at the knees of the grim
" man. A Rhadamanthine kind of man, inexorable as
" the just Gods are, desperately set upon his will here,
" and now, in the carrying of it out, pestered with *more*
" feminine interferences. Which thing a hero Grim-

"wold, intent upon ''Ware poultry' and 'Let justice be
"'done!' will nowise in the least tolerate. Lifting his
"great boot therefore—boot most iron-efficacious, as
"a beloved Bertha knows—a stout sufficient Grimwold
"conclusively *spurns* from him a foster-mother, ululat-
"ing now *too* excessively; smites her down senseless to
"the sward there—very Rhadamanthine indeed—and
"instantly, at signal given, twenty poor figures rush up
"towards the serene spaces, and, under their sufficient
"oak-bough, may kick and contort themselves at plea-
"sure, each according to his whim and private notion of
"it. Singular dance upon nothing going on here in
"the summer dawn, the opening heavens smiling down
"upon it; singular! picturesque enough! lively! natur-
"ally—each poor losel having his whim of it—with
"much convulsed variety of *step:* highly curious to be
"looked at, curious, and very edifying. 'Hideous!
"shriekest thou, O blockhead? 'Grinding of human
"'hearts under millstones!' other the like shrieking and
"fatuity. Hideous! yea truly! as the doings of the
"Gods are, which are also much other than hideous.
"Terrible we will call it; grim-tragic, which will also
"mean, well considered, grim-beautiful, and afar off
"Benign. Verily, let justice *be* done—our ''Ware
"'poultrys' and divine messages preached abroad even
"so, surely with sufficient emphasis. Truly a most
"stern man! Rhadamanthine-inexorable; with Berser-
"kir rage in him, nearly to all extents; yet, with soft

" wells of pity withal, deep down in the rude rock heart
" of him; the soft quality of mercy—when permissible,
" as hitherto it clearly could not be—nowise omitted in
" the making of him, as instantly falls to be illustrated.
" For lo! now! our score of losels, set to dance upon
" nothing there, with lively varieties in the *step* of them,
" seem shortly as if they tired of it; wax less and less
" lively, as is natural; one by one, at length wholly
" strike work, and hang there satisfactorily danced *down*,
" defunct. One most obstinate-lively losel—Foster-
" brother Wat, as I rather think—having danced down
" all the rest, still obstinately keeps dancing. The all
" too lively Wat! the singular contortings, *steps* of him
" for a time curious-amusing, now fast becoming afflic-
" tive; poor Wat, who deserved a little, getting plainly
" now *too much* of it; yet lively, and, unless we stop
" him, will go on to give himself *more*. On the whole,
" will not a Grimwold, stern, but, with wells of pity
" deep down in him, show mercy upon poor Wat *now*—
" when, perhaps, it may be permissible? Surely a
" Grimwold will show mercy, who has funds of softness
" in him withal. Wherefore, at signal again given, up
" the great oak trunk, alert as cat at it, goes *swarming*
" a deft functionary; deftly ascends; swiftly and deftly
" runs out on the sufficient oak-bough, swiftly and
" deftly, by rope, descends on poor Wat, still lively
" there; and *there*, on the poor struggling shoulders of
" him, Grimwold humming him tune for it, will per-

" form some sufficient fandango. Neck bones of Wat,
" says Jocelinus, audibly cracking under his operations.
" *Cracking!* we thank thee for the word, O Jocelinus,
" veracious human chronicler with *ears!* Whereby
" poor Wat presently, his sore sorrows now over, will
" also strike work and hang quiet like the other nine-
" teen, satisfactorily danced down and defunct. Due
" contortings for Wat, since it must be so, but not *un-*
" *due*; for the raggedest losel of them all not *undue!*
" Let justice *be* done! most sure, certain; yet also
" surely, when permissible, let mercy temper justice!
" Due contortings for Wat, not *undue*; for a poor
" Foster-brother whom we love, surely never *undue.*
" The tenderness, the fine pity of it in so grim a man as
" this is, has seemed to me, I do confess, most beautiful,
" idyllic-touching. A Grimwold surely, who has bowels
" in him, though not moving them on slight occasions.

" On the whole, can it seem other to us, than that
" this Grimwold, energetically hanging his losels here,
" let blockheads shriek as they will of it, is doing a
" manful and cheering feat under the sun? A hero
" Grimwold who, in these sad deep sunk times of 'juries
" ' declining to convict' should be very didactic indeed
" for us in this special department of things. 'Juries
" ' declining to convict!' O heavens! was ever in this
" God's world the like thing before heard of? Of *such*
" juries, what in the Gods' name is an earnest soul to
" say or think? juries which strike one DUMB as with

"awe and a certain panic terror. Hideous summary "and concrete of all practical human baseness, dastard "falsities, and stupidities whatever. Heaven send them "only a Grimwold to be didactic to them by his prompt "method of the hemp rope and sufficient oak bough. "*He*, as I do perceive, would be the right one to *reform* "such singular juries for us; *he*, and no other."

Of which highly peculiar "utterances," what is to be said except that Hero-Worship, too deep consideration of our own sublime sitting parts, and pursuit of one particular class of ideals, will be exceedingly apt, like misery, to "bring us acquainted with strange bed-"fellows." Really a hero Grimwold this to whom we must decline to bow the knee. Not an idol for *our* money this at all. With respect to reasons of dissent, of civilly declining to bow, needless, too obviously, to talk to Sauerteig. Sauerteig at this time of day got clearly beyond being talked to. A Sauerteig, who, plain Brute being presented to him, will forthwith label him "Baresark," and consider he has done the business; has as good as sprinkled holy water over him and consecrated him to all time; *such* a Sauerteig is plainly a hopeless case, and need not greatly be talked to. Else might not one feel disposed to interrogate Sauerteig a little on the head of his murderous savage and ruffian, *hight* Grimwold; Hero-Governor, whom Sauerteig so infinitely admires, teaching his dim populations to soar heavenward—twenty at a time, as we saw. As instance, not to press the case of Bertha (his *wife*, and no doubt deserving all she got and more) was his treatment of his poor foster-mother really quite Christian and humane? of foster-brother Wat, on whom he showed such singular *pity?* Rhadamanthus! responds Sauerteig, curt-taciturn; Junius Brutus! No word further from Sauerteig, except perhaps, if you still keep pressing him, "Owl! " ostrich! idiot! wholly without eye for the heroic!" Again, it might be asked, admitting all methods of justice approximate hitherto, was not the Grimwold method here a little *too merely* approximate? Did it not perhaps occur to Sauerteig, that these twenty poor losels, whose " convulsed variety of *step*" seems so edifying, amusing to him, were after all *innocent?* "Innocent!" the Sauerteig will echo, not without surprise, contempt; and, perhaps, proceed sardonically —"Who then *is* innocent? O paltry wretch! art " *thou* innocent? and if we now summarily clutched " *thee,* and, by swift Grimwold methods, throttled the " foul soul out of thee, wouldst thou then be getting " other than the God's justice, and authentically *thy* " deserts?" What to say of a Sauerteig capable of such an *argumentum ad hominem* as this? A Sauerteig who need not be talked to; who may as well without interference be left to go his own strange courses, and proceed upon his worship of Brutes by the method of labelling them Baresarks. Of his high hero Grimwold, though we, for our small part, must utterly decline the worship of him, be much joy to Sauerteig! A Sauerteig who—to show what lengths he will go—his Grimwold, by much *mead* and the like, exploding at length upon him in mere spontaneous combustions, will lovingly linger over the oleaginous-obscene deposits of him, not without questionable allusion to Elijah and fire-chariots.

Further specimens of Sauerteig we should like to give at some length, but, alas! must not. His unparalleled chapter, for instance, entitled, " Flea Hunt— Divine Significance of Fact"—could it prove other than most interesting? How a high Grimwold once at dead midnight, hero-snoring beside his beloved Bertha, dimly became conscious of sensations most itchy-uneasy on the haunch of him; flea or other vivacious insect of democratic tendencies having invaded that region, and proceeded to extract his life-fluids. How a high Grimwold woke up; swore a little, *per os Dei*—his favourite if not sole piece of piety—scratched the afflicted part, and sulkily re-addressed himself to his slumbers. How it would not in the least do; flea still most vivacious-annoying, diligently extracting the life-fluids; haunch still most itchy-uneasy; till at length an infuriated Grimwold will fairly dash out of bed imprecating heaven-high, and with much sounding of gongs, rushing of terrified lackeys with torches, (mostly in a state of entire nudity,) and other the like tumult, proceed to *hunt his flea;* beloved Bertha, in her singular night gear shivering observant the while. How, for a space of two hours, he hunts—fierce-assiduous, desperate to catch his flea; hunts, hunts, "hugest, tumultuous, in-" extinguishable Flea Hunt," says Sauerteig, "that ever " perhaps transacted itself on this God's earth;" hunts and evermore hunts, and finds, to his much rage and grief, that flea, like *vulpes* on a previous occasion, once for all, *will not be caught*—uncertain to this hour whether after all it were Flea or Bug. All this, told in the vivid Sauerteig manner, with graphic touch and due vigour of presentment, readers might have found interesting. Nay, if Sauerteig is to be believed in the matter, there is in it didactic meaning of the deeper sort. "Hugest, &c., Flea Hunt," says Sauerteig, " that ever perhaps transacted itself on this God's " earth; *which*, on the deep ground that it veritably " *did* so transact itself *there*, is precious and for ever a " possession to me. Infinite is the significance of *fact*, " of reality. Consider it, O reader; this thing actually " *was;* was, and very literally *is* now, and cannot for " ever cease to be; a portion of the God's fact which " liveth and endureth for ever. A Grimwold scratching " his haunch there, tumultuously hunting his flea there, " is great; is memorable to me; on the deep ground " that the high man *actually did it.* Demonstrable, O " reader, scientifically certain, that this very sentence I " now write is, in the turn of it, twist of it, determined,

"influenced, in infinitesimal incalculable, most name-
"less yet withal most real methods, by a Grimwold
"scratching his haunch there, in that extinct old cen-
"tury of time." We may be permitted to observe
here, that if the main function of Grimwold scratching,
be to determine the twist of the singular Sauerteig sen-
tences, the world does not perhaps on *that* head owe
any very deep debt of gratitude to Grimwold. "Flea
"or bug," proceeds the singular Sauerteig, "point
"much laboured by Dryasdust, the dim doleful creature
"that he is! with next to no result whatever for us.
"Flea or bug? question of some depth of import;
"hecatombs of human creatures burnt, martyred, mas-
"sacred to all extents, for questions, as I do perceive,
"intrinsically much more trivial; question which———."
It is not, perhaps, highly essential to follow Sauerteig in
the interesting discussion which ensues—discussion in
which Sauerteig displays his usual erudition and ability,
and flouting at ineffectual Dryasdust as he goes, conclusively establishes for all men, that once for all it was
flea—and by no means Bug, as heretical persons have
contended. In which important additional certainty,
and piece of the actual God's fact, may lie many meanings for a Sauerteig. A Sauerteig on this question of *fact*,
its divine significance and relation to thing called fiction,
not always quite easy to be made out; a little hallu-
cinated or so, perhaps; not altogether in his right mind.

Of singular chapter, entitled "Reformed Parliament,"

in which Sauerteig proposes to *hang* the universal British people, (a company of foreign artists being engaged for the occasion,) and " *so* reform it in perhaps a sufficiently " *radical* manner"—a hero-ruler, adequate to that high feat, being, at present, the one thing needed—nothing here to be remarked, except that it has suggested to us a few, perhaps rather pertinent, observations, which we take leave to entitle

HOROSCOPE.

Much meditating Sauerteig this long while, and the strange ways he is going, one wonders where he will get to in the long run—what the deuce is in the end to become of him? It is the curse, as we perceive, of this Sauerteig hitherto, that it has not lain to his hand to *do* heroisms, but only to unutterably shriek and *write* about them; course, as Sauerteig himself well knows, leading too frightfully nowhither. For Sauerteig, much dissatisfied, deeply diseased mortal, profoundly Wertherish to this hour, we observe, surprising as some may think it; a whole fierce Werther and monster brood gnawing, gnawing at his poor inwards, though the right Spartan manhood of him be nowise now minded to shriek thereof; Werther come back upon us in very singular figure, having decisively *cut* the poor sentimental and personal concern, and gone with a *will* into the hero-business; with such difference, therefore, in the aspect and practical outcome of him, as the different conditions will imply; man who will for ever fiercely curse, and hurl wild scorn at Werther, in token that he can never get wholly rid of him. For *such* a Sauerteig, what medicament save in *work*, actual hero-business to be *done*, not endlessly shrieked and written about? *Work!* which might actually be found for Sauerteig, and very much to his mind too. A high Hero-Calcraft, sole possible hero figure, and victorious *doer* in these sad times, of whom Sauerteig is in a sense the spiritual complement; Hero-Calcraft being now far spent, fordone with long life of arduous heroisms, the nerve of him much gone, as was seen in his sad bungle of the Bousfield business;* seems nowise unneedful we look about us for a fit Hero-successor of him. And does not a Sauerteig stand ready to snatch the rope from the failing hand, and victoriously

---

* This implied slur upon the character and efficiency of an eminent public functionary must now in mere fairness be withdrawn. Mr Calcraft has since, by some years of splendid professional success, entirely re-established his previous high character as a hangman; and the little difficulty which occurred with Mr Bousfield is now only remembered as one of those critical instances in which a great man has unaccountably been found beneath himself; like Napoleon on the field of Borodino, or Mr John Stuart Mill in his reasonings concerning Moral Liberty. One of Mr Calcraft's very latest efforts—his despatch of the unfortunate Dr Pritchard—the writer, as present in an official capacity, had occasion to inspect very closely; and it seemed to him the work of a master-genius in his art. Without meaning to disparage the admitted genius of Mr Carlyle, he is by no means quite sure that that gentleman —had he undertaken to "abolish the scoundrel"—would have done it very much better.

bear it forward? Would the Woods and Forests perhaps look to it? they, or Downing Street, or whoso may hold in hand the high appointment? A Sauerteig once well installed therein, duly provided with rope, and set to abolish our scoundrels for us, had we not then, for once at least, most authentically, the "right man in the right "place?" How would a Hero-Sauerteig go with his whole soul into the work, and emulate the Grimwolds whom he worships! How nicely would he handle his criminal, "using him as if he loved him!" With how grim a gusto, yet tenderly, politely withal, would he manipulate about the throat of his scoundrel; delicately trim the noose, give trimmest last touch to the nightcap, and proceed consummately to *turn him off*, a most finished and completed piece of art. A Sauerteig by whom it would almost be a happiness to be hanged; to whom surely no sufferer of proper feeling, principle, could grudge his little perquisite of the body clothes. To the public, the services of a Hero-Sauerteig would be priceless. And to Sauerteig himself—now *doing* the Hero-work, not merely shrieking and writing about, and about, and about it—surely the spiritual benefit would be much. A Sauerteig no longer isolated; haughtily, angrily aloof, as now; but more and more a man among men; who, by steady sedulous hanging of his fellow-creatures, would more and more humanly reconcile himself to them, recognise his brotherhood with all men. *Here*, we do perceive, lies the true final hero-field for

Sauerteig. Will Downing Street, when the vacancy occurs, be good enough to look very strictly to it?

Of Sauerteig why further? Of his Hero-business— mere cookery, and "the patent inimitable" nine-tenths of it—we have already seen enough perhaps. Of his "earnest soul," "noble life," and the like, what should fall to be said, except that, for souls perhaps in a small way earnest-noble, but not dreadfully *intent* upon being so, *conscious* of being so, it is really afflictive, and in fact grown to be one of the main nuisances of life in these sad times. Seems to us the "Divine meanings of "Silence" might be nowhere more obvious to Sauerteig than in this of the Earnest-noble. The Earnest-noble, shrieking itself at us from the housetops, is questionable, suspect to us. To shriek upon the housetops, O Sauerteig, really such a very easy matter; sufficient lungs of leather, we perceive, sole gift requisite for *that* exploit. In Heaven's name, O Sauerteig, *be* earnest! *be* noble! to quite infinite extents, if thou wilt, *that* being thy particular whim of it; *be;* and let it altogether suffice to thee; and the less said about it the better perhaps.

Of Igdrasil, the Life-tree again, and the highly peculiar relations of Sauerteig therewith, much might readily be said, the Time-spirits and Printer's Devils permitting. Relations surely *most* peculiar! On the whole, nothing can exceed the respect of Sauerteig for his Igdrasil; Igdrasil, which, at times, he will also lovingly denominate "the All;" or Awl is it perhaps? supreme creative cobbler's implement (strictly *without* cobbler) wherewith our great World-boot shall fashion itself and be fashioned. Igdrasil, *plus* mere Grimwolds and other the like foul Fetishes, as more and more becomes obvious, sole objects of worship, and entire spiritual furniture of the man, wherewith he will front the roaring eternities, immensities. On the whole, deepest respect, reverence for his Igdrasil; and yet, curiously enough withal, deep settled *discontent*, with an Igdrasil *growing* surely of late on palpably erroneous methods. A not quite *wise* Igdrasil, to whom Sauerteig plainly considers himself competent to give hints, wrinkles, putting Igdrasil up to a thing or two; Igdrasil whom an earnest Sauerteig will evermore correct, instruct, and *teach*, with really exemplary pains, the important lesson, *how to grow*. Not the thing at all this, as I compute, O Igdrasil! growing *now*, by these sad unexampled methods, mere *new shoots*, which are next to no good at all to us. New shoots not the thing at all, and will never do. The real thing for *thee*, O Igdrasil! to *resuscitate the dead branches* of thee; this or the other dead branch, Hero-governor or the like, rotting at the tree-root there, the old women picking it for firewood; *that*, above all, O Igdrasil, must thou resuscitate, re-inweave—begging it back from the old women—or an Igdrasil got into bad latitudes, I rather fear. Even so unutterably jargons Sauerteig, scolding, flouting at his Igdrasil, and really, with fierce pains, *teaching it how to grow*. By venerable understood methods, O Igdrasil! which I, Sauerteig, will prescribe to thee. Igdrasil, meanwhile, grows steadily, and no doubt *having* its methods as of old it had, heeds little what even a Sauerteig may think of them. In heaven's name, O Sauerteig, *let it grow*. A Sauerteig, diligently worshipping his Igdrasil, yet evermore taking to task his Igdrasil, cursing at his Igdrasil, and really with fierce pains *teaching it how to grow*, is surely an amazing spectacle for us.

Amazing; not uninstructive, significant; the Sauerteig attitude here more or less typical perhaps of some dark disunion, unreconcilement, conditioning the whole activity of the man. Man, to this hour, as we perceive, never wholly at one with himself, let him shriek and asseverate as he will of it; very "Everlasting Yea" of him, properly a *kind* of—Nay; Nay, with wild, shrieking, despairing protest against *itself*; clutching out in search of—Yea—in perhaps somewhat a blind manner, catching mere Grimwolds, Igdrasils. Yea, much worth speaking of, conclusively not to be got at, it should seem, on the questionable Sauerteig terms. Phantasms of Yea to be got at merely; wretched illusory semblances of it; wholly unsatisfying spectres of Igdrasil, Grimwold, and the like; wherewith the earnest soul, in deep just dissatisfaction withal, shrieks wildly that it *is* satisfied. Shrieks, and evermore shrieks; and much writhing, as in chronic agony and exasperation, satisfactorily testifies *so*, to what a pinnacle of superior "blessedness" it has been privileged to soar by these methods. *Blessedness*—" happiness " having been summarily kicked overboard as unworthy of us. Happiness a quite too despicable matter, unworthy the consideration of a Sauerteig; (who withal, perhaps, like another, might scream with a sufficient cramp in the belly of him.) Attitude superficially heroic; not wholly without its plausibilities, deceptive nobilities, and airs of the high old Stoic species. "Happiness unworthy of a "Sauerteig," looking to be exceeding great, and obviously so considering itself, seen to be *other* than quite great; to be more or less only *sham* great; even so far as it *is* great, to be questionable, heathenish; reconstitution on a higher plain of that very detestable Egoism, which it brags to have cast out on a lower one. Egoism, as we suspect, in some more or less damnable and deadly form of it, the inevitable outcome of Igdrasil; the Ego of which Igdrasil is an implicit suppression and outrage, avenging itself even *so*. On the whole, we surmise this Igdrasil, or Awl, to be a—Hum—for any good we are like to get of it. Horror of heart and loneliness—crushing and weary sadness, the grief which consumes and kills! *That*, we take it, is about the *net* result of Igdrasil to souls with any deep funds of natural religiosity in them; result, from which here and there a strenuous Sauerteig will with toil of heart contrive to *escape*; and realise for himself, *on his own strength*, surely a right noble and manful, if still somewhat tragical and hapless, manner of existence; an indomitable sort of Sauerteig, who will contrive in some grim-noble form to *live*, where weaker souls might sink and die, stifled in the nameless quagmires. An Igdrasil satisfactory to the intellect, deadly to the souls of men; good as intellectual conception, otherwise not quite so good; constituted into worshipable entity, found to be a cruel and ghastly idol, crushing out, as under merciless Juggernaut wheels, the hearts and lives of its worshippers. An Igdrasil, on the intellectual side, seen to be satisfactory; seen also, on the other, or moral and emotional side, imperatively to demand, for its reconcilement to the ineradicable instincts of men, recognition of some other and complementary element. Element, we suspect, quite *other* than the mere Fetish-Grimwold one; element, let us admit, in these most uncomfortably, tragically illuminated Epochs, not quite so easy to be got at, as might seem to our benighted Grandmothers. Sauerteig and the religious question! O heavens! would not an entire and prolonged Discourse, of quite *other* than the all too occasional kind, be needed for the least elucidation of so deep and perplexed a topic?

On the whole, for this Sauerteig, though at times we may do a poor snigger at him, killing our dull hour that way, we can have nothing but comparative respect; Sauerteig, though much an oddity in his way, always a high and shining figure for us. Man indeed, whom smallest blockheads may controvert, criticise; whom wise men, according to their wisdom, will be shy of trying to *instruct*. Man who indeed at times will wildly overlook much, yet who often, as from casual light-gleams, points of insight, a right reader of the cookery books may discern, sees somewhat more than he will *seem* to see; who, if looking to be dullard a little now and then, has doubtless his deep reasons for it; whom this and the other pert person, with his "scientific con-"ception of human history," may profitably pass without meddling with. Not easy, we suspect, in any of the intellectual provinces to *suggest* what should be *news* to Sauerteig, taking quietly account of much which he wildly should seem to ignore. A Sauerteig, who, the whim striking him, will ofttimes pluck the *eye* out of his *occiput*, plant it in the shining forehead of him, and *look* with really much depth and decisiveness into this and the other matter; will most pertinently now *see*: and anon, the other whim striking him, will wildly, wilfully, *not* see, and, snatching the unfortunate *eye*, stick it wildly into his *occiput* again. Truly a wild man and a wilful; luminous-tenebrific, sagacious-inept, to an extent not hitherto seen among mortals perhaps; man controvertible to nearly all extents, yet, on the whole, whom sagest persons of the discreeter sort will be shy of trying to *instruct*. O Sauerteig, high-absurd mortal that thou art! endless are the whims of thee, the humours of thee, the ground and lofty tumblings and oddities. Which of us all, inspecting the parts of thee, the curiosities quaint-absurd of thee, but continually will he, nill he, and if not *with* thee, then *at* thee, must go upon the broad grin! At the lowest, an amusing Sauerteig! Live Sauerteig! and when next he "rides abroad" on his Prose Pegasus, with surely the remarkablest paces ever exhibited by animal, "may *we* be there to see," even at that huge extortionary figure of one Pound per *vol.* for the spectacle. As a master of curious horsemanship, we consider him much beyond Gilpin. And now summarily an end of Sauerteig, and of these our all too occasional discoursings concerning him.

Ballantyne, Roberts, & Co., Printers, Edinburgh.

www.ingramcontent.com/pod-product-compliance
Lightning Source LLC
Chambersburg PA
CBHW030434190426
43202CB00036B/245